SpringerBriefs in Bioengineering

More information about this series at http://www.springer.com/series/10280

SpringerBriefs present concise summaries of cutting-edge research and practical applications across a wide spectrum of fields. Featuring compact volumes of 50 to 125 pages, the series covers a range of content from professional to academic. Typical topics might include: A timely report of state-of-the art analytical techniques, a bridge between new research results, as published in journal articles, and a contextual literature review, a snapshot of a hot or emerging topic, an in-depth case study, a presentation of core concepts that students must understand in order to make independent contributions.

Theodore A. Wilson

Respiratory Mechanics

Theodore A. Wilson
University of Minnesota
Department Aerospace Engineering
  and Mechanics
Minneapolis, MN, USA

ISSN 2193-097X   ISSN 2193-0988   (electronic)
SpringerBriefs in Bioengineering
ISBN 978-3-319-30507-3   ISBN 978-3-319-30508-0   (eBook)
DOI 10.1007/978-3-319-30508-0

Library of Congress Control Number: 2016935411

© Springer International Publishing Switzerland 2016
This work is subject to copyright. All rights are reserved by the Publisher, whether the whole or part of the material is concerned, specifically the rights of translation, reprinting, reuse of illustrations, recitation, broadcasting, reproduction on microfilms or in any other physical way, and transmission or information storage and retrieval, electronic adaptation, computer software, or by similar or dissimilar methodology now known or hereafter developed.
The use of general descriptive names, registered names, trademarks, service marks, etc. in this publication does not imply, even in the absence of a specific statement, that such names are exempt from the relevant protective laws and regulations and therefore free for general use.
The publisher, the authors and the editors are safe to assume that the advice and information in this book are believed to be true and accurate at the date of publication. Neither the publisher nor the authors or the editors give a warranty, express or implied, with respect to the material contained herein or for any errors or omissions that may have been made.

Printed on acid-free paper

This Springer imprint is published by Springer Nature
The registered company is Springer International Publishing AG Switzerland

Preface

Look back on time with kindly eyes
He doubtless did his best

Emily Dickinson

The study of the mechanics of the lung and chest wall goes back to ancient times. This study was revived in the renaissance and continued with the rise of science in subsequent centuries. The modern surge in respiratory mechanics is traced to the work of Otis and Rahn at the University of Rochester, stimulated by the drive to fly at higher altitudes and with greater accelerations during and after WWII. The field blossomed beginning in about 1960. Strong groups emerged at several institutions: Montreal led by Peter Macklem and Millic-Emili, Mayo Clinic led by Bob Hyatt, Johns Hopkins led by Sol Permutt, and, in more isolated instances, Nick Antonisen in Winnipeg, the group of anatomists led by Weibel at Bern, Jack Hildebrandt in Seattle, and surfactant scientists John Clements at UCSF and Sam Schurch in Manitoba. The foremost group was that at the Harvard School of Public Health, headed by the dean of the field, Jere Mead.

I began applying fluid mechanics to respiratory problems on my own, but by the mid-1960s, Jere Mead had taken on an engineer, and Bob Hyatt wanted to follow suit. I began collaborating with him and later with Joe Rodarte and Ken Beck at Mayo. Collaboration with Andre De Troyer began at Mayo and continued after he returned to Brussels. These people provided my education, guidance, encouragement, and scientific resources. They also introduced me to the larger community, and that community, following Jere's lead, welcomed newcomers gladly. The following 50 years have been productive and enjoyable, and I am grateful to the community in general and the group at Mayo in particular for the pleasure of spending my professional life in their company.

This monograph contains three chapters. The first describes the mechanics of the parenchyma that underlies the pressure–volume curve for uniform lung expansion and describes two important nonuniform lung deformations. The second describes the action of the respiratory intercostal muscles and the diaphragm. The third describes flow, including maximum expiratory flow, gas transport in the airways, and what is known phenomenologically at this point about nonuniformity of alveolar ventilation. It seems logical to include the pulmonary circulation and gas exchange as part of respiratory mechanics, but for some reason, perhaps simply by historical accident, the community that studied those topics was somewhat separate from the community that studied the topics covered here.

I know of no monograph on this subject, except perhaps for the volumes on mechanics in the 1986 edition of the *Handbook of Physiology: Respiration*. In the preface to those volumes, Mead and Macklem wrote that "We are still just beginning to describe breathing adequately, let alone understand it. We have only the vaguest notion as to the relative importance of tissue and surface forces in lung recoil. We appear to have no idea at all about the physiological role of smooth muscle. We have yet to agree on the actions of the respiratory muscles, and the ghost of Hamberger is back among us." Since then, the contributions of surface tension and tissue forces to lung recoil and the respiratory action of the intercostal muscles have been described. Although the physiological function of smooth muscle is still unknown, much more is now known about the properties of smooth muscle and the mechanics of constricted lungs. In addition, the source of ventilation/perfusion heterogeneity and the mechanics of the pleural space are better understood.

In the late 1980s, NIH gave ample warning that money was shifting to molecular and cellular biology. NIH is insulated from and unfettered by concerns about the welfare of people on its grant payrolls. It does not fire anyone or dictate anyone's activity; it simply takes the money from here and puts it there. Some 40–50 references are cited in each chapter of this book. Except for the book by Hamberger (1740), the dates range from 1951 to 2014 with the peak years in the 1980s and 1990s. Like symphony orchestras, newspapers, and fountain pens, the study of respiratory mechanics has disappeared from human affairs. But the physiology has not, and I hope our understanding of that physiology is preserved.

A final note on the field. Much of the data on respiratory mechanics was obtained in dogs. Now, animal rights groups have imposed restrictions on the use of dogs in research, and as a result, experiments on dogs have decreased to nearly zero. Only bred-for-research dogs that have minimal contact with humans are used. Thus, despite the fact that tens of thousands of dogs are euthanized in pounds each year, more dogs are being created to be killed intentionally. I do not see that the welfare of animals has been served.

Minneapolis, MN, USA   Theodore A. Wilson

Contents

1 Lung Mechanics 1
   1.1 Empirical Pressure-Volume Curves 1
   1.2 Parenchymal Mechanics and the Origin of Lung Recoil 2
        1.2.1 Lung Macro-Structure 3
        1.2.2 Parenchymal Micro-Mechanics 3
        1.2.3 Surfactant 5
        1.2.4 Quantitative Model 6
        1.2.5 Dissipative Processes 7
   1.3 Non-uniform Lung Deformations 8
   Appendix 14
   References 16

2 The Chest Wall and the Respiratory Pump 19
   2.1 Design of the Respiratory Pump 19
   2.2 Rib Cage and Intercostal Muscles 22
        2.2.1 Respiratory Effect of the Muscles 24
        2.2.2 Mechanisms of Intercostal Muscle Action 26
   2.3 Diaphragm 29
        2.3.1 Respiratory Effect 30
        2.3.2 Transdiaphragmatic Pressure 30
        2.3.3 Volume Dependence 32
   2.4 Other Respiratory Muscles 32
   2.5 Compartmental Models 33
   2.6 Work of Breathing 37
   2.7 Mechanics of the Pleural Space 38
   Appendix 39
   References 40

3 Flow and Gas Transport ... 43
- 3.1 The Bronchial Tree ... 43
- 3.2 Flow ... 44
  - 3.2.1 Higher Frequency Oscillatory Flows ... 46
- 3.3 Expiratory Flow Limitation ... 48
- 3.4 Convection and Diffusion ... 53
- 3.5 Ventilation Distribution ... 55
- Appendix ... 58
- References ... 59

Index ... 63

Chapter 1
Lung Mechanics

Abstract Recoil pressure in the saline-filled lung is a unique function of lung volume. This recoil is provided by forces in the tissues that form the macrostructure of the lung: the pleural membrane, the bronchial tree, and the inter-lobular membranes that connect the bronchial tree to the pleural membrane. In the air-filled lung, recoil pressure is a function of lung volume and the internal variable, surface tension, which depends on volume history. The additional recoil pressure of the air-filled lung is the result of the direct effect of surface tension and the indirect effect of inducing tension in the lines of connective tissue that form the free edges of the alveolar walls at the boundary of the lumen of the alveolar duct. Non-uniform deformations are analyzed using the methods of linear elasticity. These include the gravitational deformation of the lung and the local deformation of the parenchyma surrounding a constricted airway.

1.1 Empirical Pressure-Volume Curves

Volume and pressure are the natural variables for describing the mechanics of the lung. Transpulmonary pressure (P_{tp}) is defined as the difference between pressure at the airway opening (P_{ao}) and pressure surrounding the lungs (pleural pressure in situ or ambient pressure for excised lungs). Lung recoil pressure (P_L) is defined as the difference between alveolar pressure (P_A) and the surrounding pressure. For static conditions, $P_A = P_{ao}$ and $P_L = P_{tp}$. With flow and pressure gradients in the airways P_L and P_{tp} may be different.

Curves of the volume of air in the lungs (V_L) vs P_{tp} for quasi-static maneuvers are shown in Fig. 1.1 [1]. These curves of volume vs pressure are called pressure-volume or P-V curves. Trajectories for different maneuvers are shown. The right-hand-curve shows inflation from a low lung volume to total lung capacity (TLC), defined as V_L at $P_{tp} = 25$ cm H_2O. The left-hand-curve shows deflation from TLC to negative values of P_{tp}. At negative values of P_{tp}, the air that remains in the lung due to trapping by airway closure is denoted residual volume (RV). Inflations to TLC from two points along the deflation limb are also shown. Tidal breathing with a tidal volume (V_T) of 1 L is shown by the line with double arrows. The volume at

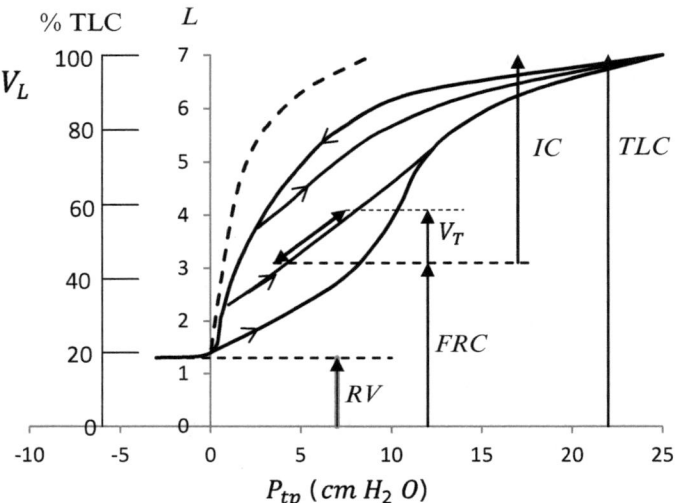

Fig. 1.1 Volume of air in the lung (V_L) in liters (L) and % total lung capacity (*TLC*) vs. transpulmonary pressure (P_{tp}) for various maneuvers described in the text. The pressure-volume curve for the saline filled lung is shown by the dashed line

end-expiration on this line is the equilibrium volume of the passive chest wall and lung and is denoted functional residual capacity (*FRC*). The elastance of the lung along the tidal breathing line is ~4 cm H_2O/L [2]. Inspiratory capacity (*IC*) is the difference between *TLC* and *FRC*.

1.2 Parenchymal Mechanics and the Origin of Lung Recoil

A schematic diagram of the pleural membrane and the attached parenchyma is shown in Fig. 1.2. In this free body diagram, the forces carried by material elements that cross the dashed line must be included in the force analysis for the segment. The magnitude of the effective stress due to these forces is obtained from the force balance for the segment. Alveolar gas pressure (P_A) acts to the right on the membrane and pleural pressure (P_{pl}) acts to the left. The force balance requires that the tissue stress equals $P_A - P_{pl} = P_{tp}$. The understanding of the mechanics of the pressure-volume curve therefore lies in understanding the forces carried by the material elements in the lung.

By de-gassing the lung and then inflating the lung with saline, the air-liquid interface and surface tension (γ) and the forces induced in tissue by surface tension are abolished. The pressure-volume curve for the saline-filled lungs is shown by the dashed line in Fig. 1.1. This leads to the concept that P_{tp} is the sum of the recoil

1.2 Parenchymal Mechanics and the Origin of Lung Recoil

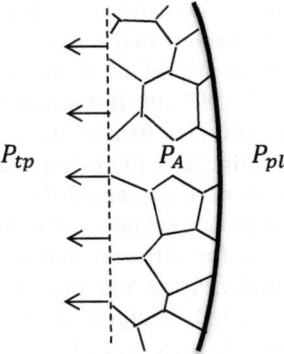

Fig. 1.2 Free body diagram for a segment of lung and pleural membrane

pressure of the saline-filled lung (P_{sal}) and an additional recoil generated by surface tension (P_γ).

$$P_{tp} = P_{sal} + P_\gamma \qquad (1.1)$$

1.2.1 Lung Macro-Structure

The recoil of the saline-filled lung is provided by the forces in the tissue macro-structure of the lungs which are unaffected by surface tension. This macro-structure consists of the pleural and inter-lobar membranes, the bronchial tree, and the inter-lobular membranes that connect the bronchial tree to the surrounding membranes. The tension in the pleural membrane has been measured [3], and the elastance of the bronchial tree is known [4], but the anatomy and mechanics of the inter-lobular membrane network is largely unexplored [5]. This network must be fairly comprehensive and complete because the shape of the saline-filled lobe, including the sharp edges and reverse curvatures of the caudal surfaces of the lower lobes, is the same as that of the air-filled lobe.

1.2.2 Parenchymal Micro-Mechanics

The acinus consists of 3–5 generations of alveolar ducts and sacks that extend from a terminal bronchiole [6]. The lumen of the alveolar duct is surrounded by polyhedral alveoli with diameters of ~200 μ. The free edges of the alveolar walls at the opening to the duct are formed by bands or cables of elastin and collagen. Some of these originate at the distal end of the terminal bronchiole forming intersecting helices of different handedness around the circumference of the lumen. The volume of the acinus is ~10^{-2} cm^3, and the total alveolar surface area of the lung is ~100 m^2, comparable to the area of one half of a tennis court. Surface tension on this surface area is one of the main contributions to P_γ.

In 1979, the Berne anatomists introduced a new method of fixing lungs [7, 8]. Previously, the fixative was delivered via the airways. The new method improved the fixative and more important, delivered the fixative via the pulmonary vascular system. This method allowed the observation of qualitative features of parenchymal architecture under the influence of surface tension, and it provided data on surface area for different states of saline- and air-filled lungs. The Berne anatomists analyzed micrographs of slices of parenchyma from saline-filled, air-filled, and air-filled, detergent-washed, rabbit lungs fixed by vascular perfusion and found that surface area depends on both lung volume and surface tension. In the saline-filled lungs, with no surface tension, the alveolar walls are undulating and appear to be flaccid. In the air-filled lungs, the alveolar walls are straight and excess wall material is folded into the corners where alveolar walls intersect.

These observations led to the following conceptual model of parenchymal mechanics [9]. Tension in the tissue of the alveolar walls is negligible; alveolar walls serve only as platforms for surface tension at the air-liquid interface. Surface tension exerts an outward force on the cables that form the free edge of the alveolar opening. The cables are stretched until the tension (τ) in the cables provides a hoop stress that balances the outward pull of surface tension on the two faces of the alveolar wall. This equilibrium is illustrated in Fig. 1.3. At equilibrium, $\tau/\rho = 2 \cdot \gamma$, where ρ is the radius of curvature of the cable. The tension in the cables provides a second contribution to P_γ.

Other observations are consistent with this conceptual model. First, the distribution of the angles of intersection of alveolar walls is tightly concentrated around an angle of 120°, as one would expect if the forces carried by the walls were dominated by surface tension [10]. Second, 80 % of the connective tissue in the parenchyma is concentrated in the cables that form the free edges of alveolar walls or appear at sharp bends in alveolar walls [11]. The walls contain only thin strands of connective tissue. To be sure, because of the curvature of the air-liquid interface at the alveolar corners, the pressure in the corners is lower than the pressure in the flats by γ/r where r is the radius of curvature of the interface, and this pressure difference must be balanced by a tension of $\gamma \cdot t/r$ in the wall, where t is the wall thickness, but for t/r small compared to 1, this tension is small compared to 2γ from the interfaces on the two sides of the wall.

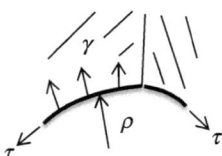

Fig. 1.3 Curved elastic cables (*dark lines*) at the free edges of two intersecting alveolar walls. A third alveolar wall and cable segment meet the line of intersection at the back

1.2.3 Surfactant

It can be seen from Fig. 1.1 that V_L is not a unique function of P_{tp}; any point within the outer lines can be reached by a particular volume history. Thus, the state of the lung is determined by an additional internal variable; namely, surface tension.

In 1955, Pattle re-discovered that lavage liquid recovered from the lung has a low surface tension [12]. Clements identified the surfactant in lung lavage fluid and found that surface tension in the fluid was strongly dependent on surface area [13]. Clements, Schurch and colleagues have made extensive studies of the physical chemistry of lung surfactant [14–16].

The alveolar walls are covered by a liquid subphase. Type II cells in the alveolar wall excrete surfactant into the subphase, and surfactant molecules are adsorbed into the air-liquid interface. At equilibrium, with a saturated subphase, surface tension is ~28 dyn/cm. If a surface in equilibrium expands, surface tension rises, but surfactant is relatively rapidly adsorbed, and surface tension returns to its equilibrium value. If surface area is compressed, the area density of molecules at the interface increases, surface tension decreases, and surface tension is stable, down to ~2 dyn/cm. With further area compression, the surface layer buckles, surfactant is ejected into the subphase, and surface density and surface tension remain constant. In the range, $2 < \gamma < 28$, the surface acts as an elastic sheet with a large specific elastance [16]. For a 5 % increase in surface area, surface tension increases by 10 dyn/cm.

If the compressed area is held constant for longer times, surfactant slowly enters the subphase, surface density decreases, and surface tension increases at a rate of ~2 dyn/cm per minute [14]. Minimum surface tension can be restored either by compressing the area further or by expanding the area so that additional surfactant is absorbed into the interface, and then compressing surface area. In subjects whose rib cage and abdomen are strapped so that they breath over a restricted low lung volume range, recoil pressure and lung elastance gradually rise, and a large breath is required to re-establish normal recoil. As Jere Mead has said, "An occasional sigh has more than emotional significance."

In 1959, Avery and Mead reported that surface tension was high in premature infants suffering from respiratory distress syndrome [17]. With abnormal surface tension, the architecture of the parenchyma is distorted, the fluid balance of the lung is disturbed and lung inflammation occurs. The surfactant supply develops late in pregnancy and is underdeveloped in premature infants. These observations have led to the development of surfactant replacement therapy that has saved the lives of many premature infants.

1.2.4 Quantitative Model

According to the conceptual model described above, P_γ is provided by two forces exerted across the plane marked by the dashed line in Fig. 1.2, surface tension on the surfaces and tissue tension in the cables that intersect the plane. The quantitative description of these two components of P_γ is given by Eq. (1.2).

$$P_\gamma = \frac{2}{3} \cdot \frac{\gamma \cdot S}{V_L} + \frac{1}{3} \cdot \frac{\tau \cdot L}{(V_L + V_{ti})} \qquad (1.2)$$

The first term in Eq. (1.2) describes the direct effect of surface tension. For randomly oriented surfaces carrying surface tension γ, the normal force per unit area acting across a plane is given by this term, where S/V_L is the surface area (S) per unit air volume [18]. For randomly oriented line elements carrying tension τ, the normal force per unit area acting across a plane is given by the second term, where $L/(V_L + V_{ti})$ is the length of the line elements (L) divided by total volume, air plus tissue volume (V_{ti}) [9]. Thus, the stress in the parenchyma is the sum of the direct effect of surface tension and an indirect effect: tension in the line elements that is induced by surface tension.

A complete mathematical model corresponding to the conceptual model described above can be constructed (Appendix). The values of P_{tp} and S predicted from this model for volume trajectories like those shown in Fig. 1.1 are shown in Fig. 1.4. The predicted pressure-volume curves shown in the left panel are like those shown in Fig. 1.1 for $V_L < 80\%$ TLC. For the deflation and inflation limbs on the left and right sides of this figure, surface tension is constant at 2 and 28 dyn/cm, respectively, and P_{tp} is nearly independent of V_L [19]. Surface tension increases with increasing V_L along the inflation limbs with lower slope.

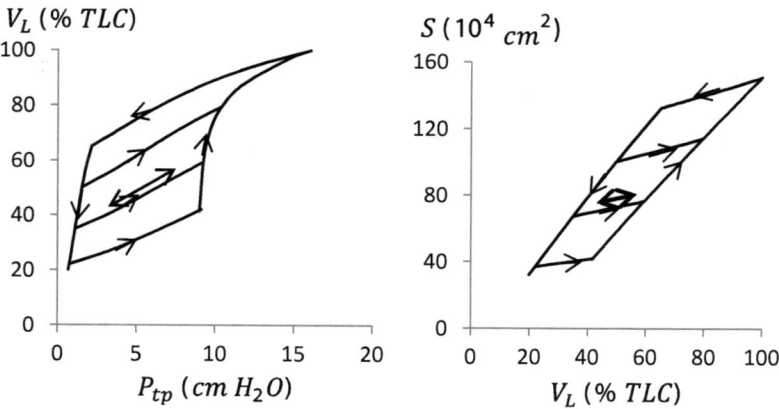

Fig. 1.4 Pressure-volume curves (*left panel*) and surface area-volume curves (*right panel*) predicted by the model

It is clear from the right panel of Fig. 1.4 that parenchymal geometry does not remain similar as lung volume changes. For the limbs with constant surface tension on the left and right sides of this panel, cable length and lumen diameter remain approximately constant as lung volume changes. As a result, S decreases or increases more rapidly than $V_L^{2/3}$. For the inflation limbs along which γ is increasing with increasing V_L (lines with smaller slopes), cable length increases and subtends an increasing fraction of duct volume. For these segments, S increases less rapidly than $V_L^{2/3}$. These different relations between S and V_L are like those observed [7, 8].

The model does not represent the increased elastance at higher lung volumes shown in Fig. 1.1. Perhaps at higher volumes, the slack in the alveolar walls is taken up and tension in the tissue contributes to recoil. If so, the parenchymal geometry may depart from the geometry that minimizes surface area; namely, flat walls and intersection angles of 120°, and surface area increases more steeply with lung volume.

1.2.5 Dissipative Processes

For tidal breathing, the pressure-volume trajectory for expiration does not retrace the trajectory for inspiration. Pressure during inspiration at a given volume is higher than that during expiration and the trajectory for a breath forms a loop with the area of the loop equal to the energy dissipated during the cycle. Two dissipative processes contribute to the loop.

The first is viscous dissipation in the flow through the airways (Chap. 3). The second is tissue viscance. This is measured by the area of the pressure-volume loop for low frequency volume oscillations or by the area of the alveolar pressure-volume loop. For tidal volumes of <25 % *TLC*, the areas of these loops are found to be independent of frequency [20] and proportional to V_T^2 [21]. For sinusoidal volume oscillations, this process is modeled by a resistance with a magnitude that is inversely proportional to frequency. This model is called the "structural damping" model, because a similar behavior is found in structures composed of many components, or the "constant phase" model, because with resistance inversely proportional to frequency, the phase angle between the pressure and volume oscillations is independent of frequency. Tissue viscance also manifests itself in the pressure relaxation that occurs when lung volume is held constant after a rapid increase or decrease [22, 23]. The mechanisms that underlie tissue viscance are unknown.

At normal breathing frequencies and tidal volumes, tissue resistance is somewhat larger than airway resistance, the total resistance is ~2 cm $H_2O/(L/s)$, and volume lags pressure by ~35°. For tidal volumes >25 % *TLC*, surface tension hysteresis becomes significant.

1.3 Non-uniform Lung Deformations

In the discussion of the pressure-volume curve above, it has been implicitly assumed that pleural pressure is uniform, that the properties of the parenchyma are uniform (the same at every point) and isotropic (the same in all directions), that structures embedded in the parenchyma such as airways and large blood vessels expand similarly to the lung, and therefore that the strains in the parenchyma are uniform and isotropic. In fact, these assumptions are only approximately satisfied, and lung deformations are, to some extent, non-uniform.

Two non-uniform deformations are of particular interest. The first is the gravitational deformation of the lung within the chest wall. This problem is illustrated by the simplified model shown in Fig. 1.5. The weight of the lung must be supported, and therefore, pleural pressure must be more positive (less negative) at the base of the lung than at the apex. The solid lines in Fig. 1.5 mark equi-spaced planes in the uniformly expanded lung, and the dashed lines mark the locations of those planes in the gravitational field. The upper and lower surfaces are constrained to remain in contact with the surrounding chest wall, but the body of the lung slides along the chest wall, and the downward displacement from the uniformly expanded position is maximum at mid-lung height. The upper half of the lung is expanded and the lower half is compressed. Pleural pressure at the bottom of the lung is higher than that at the top by $\rho_L \cdot g \cdot h$ where ρ_L is lung density and h is the height of the lung. Thus, P_{tp} and the tensile stress in the lung tissue decrease from top to bottom, and the gravitational body force on each lung layer is balanced by the difference between the greater tensile stress acting on the upper surface and the smaller tensile stress acting on the lower surface of the layer. The gravitational gradient in regional lung volume, as a fraction of volume at TLC, has been measured in dogs [24, 25] and in humans [26, 27].

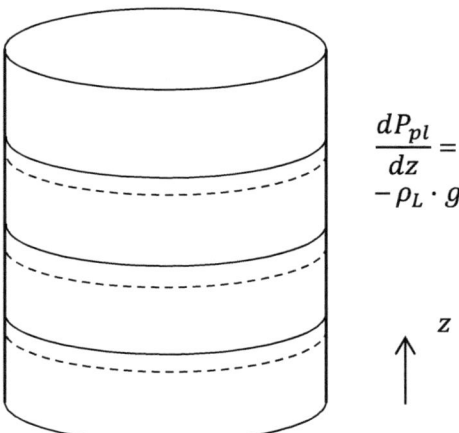

Fig. 1.5 Schematic model of the gravitational distortion of the lung in the chest wall

$$\frac{dP_{pl}}{dz} = -\rho_L \cdot g$$

1.3 Non-uniform Lung Deformations

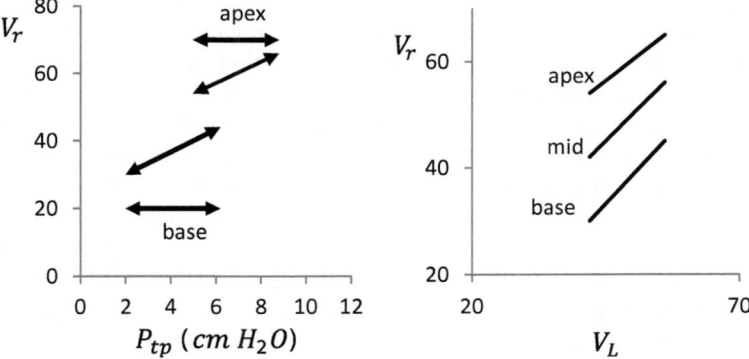

Fig. 1.6 *Left panel*: ranges of P_{tp} and regional volume (V_r) trajectories for the apex and base. *Right panel*: V_r vs. V_L, both as % TLC, for the apex, mid-lung, and base

The effect of the gravitational distortion of the lung on regional volumes is described in Fig. 1.6. For a cylinder with height 20 cm, diameter 14 cm, $V_L = 3.1$ L, and a lung mass of 1 kg, the pleural pressure gradient is ~ 0.2 cm H_2O/cm [28, 29]. Thus, if pleural pressure is -4 cm H_2O at mid-height, pleural pressure is -6 at the apex and -2 cm H_2O at the base. These pressures are shown by the left ends of the horizontal arrows in Fig. 1.6. The shape of the chest wall is assumed to remain similar as lung volume changes. For an increase in lung volume of 1 L, and $E_L = 4$ cm H_2O/L, pleural pressure at mid-height is -8 cm H_2O. The height of the lung is larger, but the density is smaller, and the pleural pressure gradient is smaller. Pleural pressure is -9.8 at the apex and -6.2 cm H_2O at the base. These pressures are shown by the right ends of the horizontal arrows. Lung elastance is roughly independent of volume [30], and the regional volume excursions are shown by the slanted lines.

Plots of regional lung volume vs. V_L as % *TLC* are shown in the right panel of Fig. 1.6. The regional volume trajectories are nearly parallel, and the gradient of specific ventilation, increasing by a factor of 2 from apex to the base, is primarily the result of the gradient of end-expiratory volumes. Milic-Emili et al. [28] measured regional volume as a function of lung volume by measuring the radiation intensity at different lung heights during wash-in of a radioactive gas. Their results show regional volume vs. lung volume curves much like those shown in Fig. 1.6.

For different postures, different parts of the lung are lower in the gravitational field. Lower regions are termed "dependent," and higher regions are termed "nondependent." In the supine posture, a wedge of lung lies below the dome of the diaphragm and part of the weight of the heart and upper abdomen is carried by the lung [31], and this adds to the pleural pressure gradient. In the prone posture, the dome of the diaphragm and the heart lie closer to the ventral rib cage, the rib cage carries part of their weight, and the gradient of pleural pressure is smaller

[29, 32]. Other studies of wash-ins of test gases show that specific ventilation is greater in dependent regions than in nondependent regions [28, 33–35].

In order to explain his observations on concentration vs. expired volume after a single breath of a test gas, Anthonison et al. [33] hypothesized that the lines shown in Fig. 1.6 are curved: concave downward for the apical regions and concave upward for the basal regions. The nest of these hypothetical curves for regions at different heights in the lung was later termed the "onion-skin" diagram. To be sure, if lung volume is extended upward so that the volumes of apical regions are greater than 80 % *TLC* where lung elastance increases, the lines converge. If lung volume is extended downward so that the basal regions approach 20 % *TLC* where airway closure occurs, the lines also converge toward a common value. However, for the bulk of the range of lung volumes near mid-volume, no mechanism that would cause curvature of the regional volume trajectories has been proposed, and no direct evidence of curvature has been reported.

The second important question concerning non-uniform deformation of the lung is the question of the stress that is applied to blood vessels and airways embedded in the lung. Mead et al. [36] introduced this question and addressed it by analyzing a two-dimensional model shown in Fig. 1.7. The parenchyma is represented by a hexagonal network of springs surrounding a cylindrical airway. In the reference state, the spring network is uniformly expanded by forces applied at the boundary. The deformed state is generated by a decrease in the diameter of the airway. Points on the springs at the boundary are fixed and points attached to the airway move with the airway and the equilibrium configuration of the network is calculated for these boundary conditions. This model illustrates several features of the mechanics of non-uniform deformations of the lung around airways. First, in the reference state, the average force per unit area exerted on a plane through the network equals P_{tp}, and in this state, the pressure exerted on the outer boundary of the airway by the

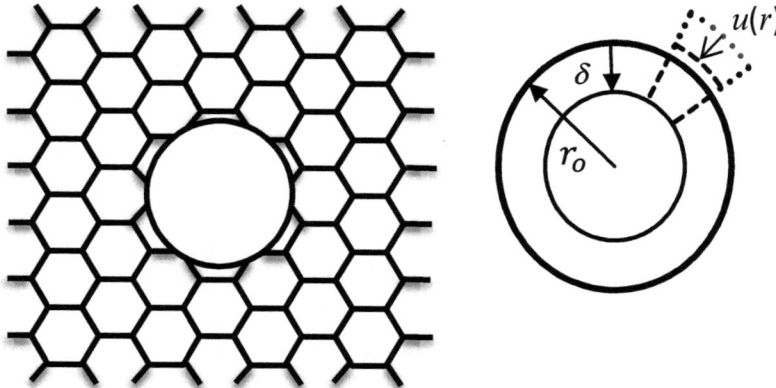

Fig. 1.7 *Left panel*: hexagonal network model of the parenchyma with attachments to an embedded airway. *Right panel*: linear elasticity model of the deformation of the parenchyma surrounding a compressed airway

1.3 Non-uniform Lung Deformations

parenchymal attachments is a tensile stress equal to P_{tp}. Under static conditions, gas pressure in the parenchyma acts as a compressive stress on the outer boundary of the airway, and the net stress on the outer boundary is P_{pl}. Gas pressure in the lumen of the airway equals gas pressure in the alveoli and the net distending pressure is P_{tp}. Thus, the airway is exposed to the same distending pressure as the lung. During expiratory flow, gas pressure in the lumen is less than alveolar pressure because of the viscous pressure loss between the alveoli and the airway, the airway is compressed and the surrounding parenchyma deformed. The hexagonal network model exhibits three potential mechanisms for an increase in the stress exerted by the parenchyma on the airway: the density of attachments of the parenchyma increases because the circumference of the airway decreases, the springs representing alveolar walls rotate into more radial orientations, and the springs that are radially oriented lengthen, increasing tension in these springs. Mead et al. described the stresses that accompany deformations of the parenchyma from the uniformly expanded state as "interdependence" effects.

A more effective way of analyzing this lung deformation is to draw on the theory of linear elastic materials [37]. If the length scale of the deformation (in this case, airway diameter) is large compared to the alveolar diameter, the forces and displacements can be averaged and the deformations can be described by the average strains and stresses in the parenchyma. With this modeling, the lore of engineering mechanics can be brought to bear on problems of lung deformations.

All small deformations of a material consist of two fundamental deformations, a dilatation or change of volume and a shear or change of the angle between lines in the material with no change in volume. These fundamental deformations are illustrated in Fig. 1.8. The stiffnesses of the material for these two deformations are described by two material properties, the bulk modulus K and the shear modulus μ. For a pure volume dilatation, $\sigma_{xx} = \sigma_{yy} = \sigma_{zz} = K \Delta V / V_o$, where ΔV is the change in volume and V_o is the original volume. For a shear deformation,

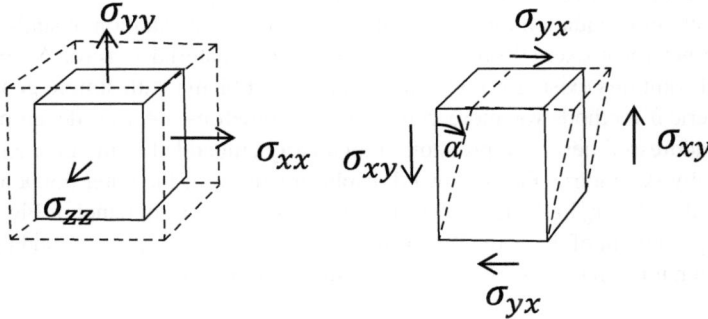

Fig. 1.8 *Left panel*: volume dilatation of a cube of material generated by normal stresses acting on the faces of the cube. *Right panel*: shear deformation generated by shear stresses acting in directions tangent to the faces of the cube. Note that the shear deformation, viewed from an $x'-y'$ coordinate system that is rotated counter-clockwise by 45° from the $x-y$ coordinates would appear to be an extension in the x' direction and a compression in the y' direction

$\sigma_{xy} = \sigma_{yx} = \mu \cdot \alpha$, where alpha is the change in angle between orthogonal lines in the reference state, as shown in Fig. 1.8.

The method of applying the linear elasticity model is the following. For a given V_L and mean P_{tp}, the uniformly expanded lung is taken as the reference state, and deformations from that state are described by stresses and strains as functions of position. The material properties depend on the reference state. The bulk modulus is related to lung elasticity by the relation, $\mathbf{K} = V_L \cdot E_L$. Thus, the value of $\mathbf{K}$ can be obtained from the slope of the pressure-volume curve for small volume displacements around the reference state. In the example above, at FRC, $\mathbf{K} = 12.4$ cm H_2O. The value of μ can be obtained by applying a standard engineering test, the punch indentation test at the surface of the lung [30]. The results show that μ is dependent only on P_{tp}: $\mu = 0.7 \cdot P_{tp}$. In the example above, at FRC, $\mu = 2.8$ cm H_2O.

Some general statements about lung deformations can be obtained from this information about $\mathbf{K}$ and μ. First, μ is considerably smaller than $\mathbf{K}$; the lung changes shape more easily than it changes volume. Second, the proportionality between μ and P_{tp} shows that the shear stiffness of parenchyma depends on the pre-stress in the parenchyma; for $P_{tp} = 0$, the shape of the parenchyma is undetermined. This is also a property of the hexagonal network model. Third, the structure of parenchyma is similar to the structure of a foam, and the linear dependence of μ on P_{tp} is similar to the linear relation between the value of μ for a foam and the pressure in the gas trapped in a foam. However, the coefficient in that relation for parenchyma, 0.7, is higher than the coefficient for foam, 0.25 [38]. Apparently the constraints on the deformation of parenchyma that are imposed by material connections at the intersections of alveolar walls and by the cables at the alveolar openings limit the micro-displacements in response to shear strains. Finally, the linear dependence on P_{tp} indicates that only the first two mechanisms mentioned above as the source of the shear stress are valid. That is, the shear stress is a result of changes in the densities of surface and line elements and changes in the orientations of these elements; changes in the forces carried by the elements are negligible [39].

The governing equation for material deformations is the equation of static equilibrium: the net force exerted on every element in the material is zero. A lexicon of analytical solutions exists for boundary condition problems with simple geometries, and numerical methods for more complicated geometries are well-developed. The problem addressed here, compression of an airway embedded in the parenchyma, is cylindrically symmetric. For an inward displacement (δ) of the outer boundary of an airway with radius r_o in the reference state as shown in the right panel of Fig. 1.7, the radial displacement of the parenchyma at the airway wall, $u_r(r_O) = -\delta$, and the radial and azimuthal normal stresses, σ_{rr} and $\sigma_{\theta\theta}$, are the following.

$$u_r = -\delta \cdot r_O/r \quad \sigma_{rr} = -\sigma_{\theta\theta} = 2 \cdot \mu \cdot \delta \cdot r_O/r^2 \qquad (1.3)$$

The displacement falls off with $1/r$, and the stresses fall off with $1/r^2$.

1.3 Non-uniform Lung Deformations

A segment of parenchyma near the airway is shown in the reference state by the light dashed lines and in the deformed state by heavy dashed lines in Fig. 1.7. The parenchyma is stretched in the radial direction, compressed in the azimuthal direction, and the volume remains constant. Thus, the stresses depend on μ only. The interdependence stress on the outer wall of the airway ($r = r_o$) is $2 \cdot \mu \cdot \delta/r_o$.

For airway compression due to the pressure drop in the airways during expiratory flow, interdependence stresses have little effect on airway caliber and flow resistance, even for maximum expiratory flow. At higher lung volumes at which the values of μ are higher, airway compression at the flow limiting site is modest (Chap. 3), and at lower lung volumes, μ is smaller and the stresses are small compared to airway elastance.

However, interdependence plays a significant role in the mechanics of constricted airways. Smooth muscle bundles lie along intersecting counter-rotating helical lines near the inner edge of the bronchial wall. The advancement angle of the helices is small [40] and the muscle can be reasonably modeled as a circumferentially oriented layer. The force (F) vs length (L) curve of maximally activated smooth muscle is like that of skeletal muscle with F decreasing nearly linearly as L decreases below optimal length (Lo). However, the intercept along the abscissa where F reaches zero is $L/Lo = 0.1$, a considerably smaller value of L/Lo than that for skeletal muscle [41]. In a cross-sectional view of a maximally expanded airway ($L = Lo$), the airway wall occupies ~16 % of the area inside the outer boundary, and the submucosa inside the muscle layer occupies ~4 % [42]. Therefore, for $L/Lo = 0.2$, tissue fills the airway and the lumen is closed, but the muscle can carry a force that exerts a compressive stress of 30 cm H_2O [43]. The radius of the outer boundary is 40 % of its expanded radius (r_o), and the displacement of the outer boundary, $\delta = -0.6\, r_o$. For the large displacement in this case, the interdependence stress is twice P_{tp} [44]. Thus, for $P_{tp} < 10$ cm H_2O, the compressive stress is larger than the parenchymal stress at the outer surface and the airway is closed, and for $P_{tp} > 10$ cm H_2O, the airway is open with the interdependence stress providing 2/3 of the opening stress. Experiments on pressure transmission between the airway opening and the periphery of constricted lungs confirm that airway opening occurs at $P_{tp} = 10 - 12$ cm H_2O [45]. In asthma, muscle hypertrophy and airway inflammation may cause the critical value of P_{tp} to be higher.

The interdependence stress depends on the level of expansion of the parenchyma surrounding an airway, and with ventilation, the expansion of the parenchyma depends on the flow through the airway and hence, on airway caliber. This connection between the size of the airway and the forces that determine the size leads to two stable states for the airways in constricted lungs [46]. In one state, the airway is nearly fully open, and flow and parenchymal expansion maintain the expanded airway. In the other, the airway is nearly closed, and the surrounding parenchyma is not well-ventilated. This bi-stable state of the airways is the cause of the bimodal distribution of $\dot{V}_A/\dot{Q}$ in constricted lungs and asthmatics. One mode is centered on a nearly normal value of $\dot{V}_A/\dot{Q}$, and the other is centered on a value of $\dot{V}_A/\dot{Q}$ that is about 1/10 the normal value.

Interdependence also plays an important role in the complicated relation between blood vessel diameter and V_L, P_{tp} and vascular pressure [44].

It should be noted that in the gravitational deformation of the lung, described above, shear deformation, as well as volume deformations occur, and this was ignored in the analysis. Because the shear modulus is small compared to the bulk modulus, the approximate analysis is reasonably accurate. A more thorough numerical analysis of the gravitational deformation for a more realistic lung and chest wall shape yields results that are much like those described above [47].

Appendix

The fundamental equation of the model is Eq. (1.2).

$$P_\gamma = \frac{2}{3} \cdot \frac{\gamma \cdot S}{V_L} + \frac{1}{3} \cdot \frac{\tau \cdot L}{(V_L + V_{ti})} \qquad (1.4)$$

Surface area increases with increasing V_L and decreases with increasing L. Thus, S is a function of these two variables.

$$S = S(V_L, L) \qquad (1.5)$$

This function is subject to two conditions. The first is the condition of internal equilibrium. At equilibrium, the internal stored energy of a structure is minimum, and the change of internal stored energy for a virtual displacement of the system is zero. In this case, the internal energy is the sum of the surface energy and the stored energy in the cables, and the equilibrium condition for a virtual change in L is the following.

$$\gamma \cdot \frac{\partial S}{\partial L} + \tau = 0 \qquad (1.6)$$

Equation (1.6) is a generalized form of the equilibrium between the outward pull of surface tension and the inward Laplace force exerted by the cable. The second condition is the equality between the pressure-volume work done by lung inflation and the increase in the stored energy in the structure.

$$P_\gamma = \gamma \cdot \frac{\partial S}{\partial V_L} \qquad (1.7)$$

Substituting from Eqs. (1.6 and 1.7) into Eq. (1.4) for τ and P_γ yields the following partial differential equation for S.

Appendix

$$\frac{\partial S}{\partial V_L} - \frac{2}{3}\frac{S}{V_L} + \frac{1}{3}\frac{L}{(V_L+V_{ti})}\cdot\frac{\partial S}{\partial L} = 0 \tag{1.8}$$

The general solution to this equation is the following, where F is an arbitrary function of its argument x.

$$S = V_L^{2/3}\cdot F\left[\frac{L}{(V_L+V_{ti})^{1/3}}\right] \tag{1.9}$$

Here, we take $F = C_1[1 - C_2x^2]$, and S is the following.

$$S = C_1\cdot V_L^{2/3}\cdot\left[1 - C_2\cdot\frac{L^2}{(V_L+V_{ti})^{2/3}}\right] \tag{1.10}$$

The first term in Eq. (1.10) describes the relation between surface area and volume for a volume change in which the geometry of the parenchyma remains similar. The second describes the deficit in surface area because of the volume occupied by the lumen of the duct. With this expression for S, Eq. (1.6) is the following.

$$2c_1c_2\frac{\gamma\cdot V_L^{2/3}\cdot L}{(V_L+V_{ti})^{2/3}} = \tau \tag{1.11}$$

Finally, the length-tension relation for the cables is given by the following empirical equation, where L_0 is the resting length of the cable.

$$\tau = C_3\left(L/L_0 - 1\right)\cdot \exp\left[C_4\left(L/L_0 - 1\right)^2\right] \tag{1.12}$$

The parameter L_0 can be absorbed into the parameters of the equations, Eqs. (1.4, 1.10, 1.11 and 1.12), by introducing the following normalized variables: $v = V/V_o$, $s = S/V_0$, $t = L_0\cdot\tau/V_0$, and $l = L/L_0$, where the reference volume, V_0, is taken as *TLC*. In terms of these variables, the governing equations, Eqs. (1.4, 1.10, 1.11 and 1.12), are the following.

$$P_\gamma = \frac{2}{3}\frac{\gamma\cdot s}{v_L} + \frac{1}{3}\frac{t\cdot l}{(v+v_{ti})} \tag{1.13}$$

$$s = c_1 v_L^{2/3}\left[1 - c_2\frac{l^2}{(v_L+v_{ti})^{2/3}}\right] \tag{1.14}$$

$$2c_1c_2\frac{\gamma\cdot v_L^{2/3}\cdot l}{(v_L+v_{ti})^{2/3}} = t \tag{1.15}$$

$$t = c_3(l-1) \cdot \exp\left[c_4(l-1)^2\right] \tag{1.16}$$

For given values of γ and v_L, Eqs. (1.13–1.16) can be solved for the values of the remaining variables, l, t, s, and P_γ. To follow a volume history, starting from given values of γ and v_L, the equation governing the evolution of γ is added.

$$s \cdot \frac{\partial \gamma}{\partial s} = 200 \text{ dyn/cm} \tag{1.17}$$

The parameter c_1 in these equations has units of cm^{-1}; its value is inversely proportional to the diameter of the alveolus. c_2 is dimensionless; it describes the fraction of the volume of the duct that is subtended by the lumen. The parameter c_3 has units of pressure; it sets the magnitude of the second term in Eq. (1.13), the pressure provided by the cables. c_4 is dimensionless; it describes the nonlinearity of the length-tension curve of the cables.

The following values of the parameters were assigned.

$$c_1 = 400 \text{ cm}^{-1} \quad c_2 = 0.3 \quad c_3 = 16 \text{ cm H}_2\text{O}$$
$$c_4 = 4 \quad\quad\quad v_{ti} = 0.12$$

The equations were solved for volume trajectories like those shown in Fig. 1.1. The outer inflation curve begins at $v_L = 0.22$, $\gamma = 2$ dyn/cm, and follows the variables as v_L increases to the value at which $\gamma = 28$ dyn/cm. It continues, holding γ constant and ignoring Eq. (1.17), to $v_L = 1$. The deflation limb follows the variables to the value of v_L at which $\gamma = 2$ dyn/cm, and continues, holding γ constant, down to $v_L = 0.2$. Inflation limbs starting at $v_L = 0.35$ and $v_L = 0.5$ and $\gamma = 2$ dyn/cm and going to the values of v_L at which $\gamma = 28$ dyn/cm were also calculated. The predicted values of P_{tp} and S for these trajectories are shown in Fig. 1.4.

It is interesting to note that the values of γ and P_{tp} are the same for mammals of all sizes, but that alveolar diameter varies over a range of a factor of 4 with diameter roughly correlated with animal mass [48]. A larger value of c_1 and a smaller value of c_3 would be used to represent a species with smaller alveolar diameter, and the first term in Eq. (1.13) would be relatively bigger than the second.

References

1. Salmon RB, Primiano FP, Saidel GM, Niewoehner DE. Human lung pressure-volume relationships: alveolar collapse and airway closure. J Appl Physiol. 1981;51:353–62.
2. Rodarte JR, Noredin G, Miller C, Brusasco V, Pellegrino R. Lung elastic recoil during breathing at increased lung volume. J Appl Physiol. 1999;87:1491–5.

References

3. Hajji MA, Wilson TA, Lai-Fook SJ. Improved measurements of shear modulus and pleural membrane tension of the lung. J Appl Physiol. 1979;47:175–81.
4. Smith JC, Butler JP, Hoppin Jr FG. Contribution of tree structures to lung elastic recoil. J Appl Physiol. 1984;57:1422–9.
5. Verbeken EK, Cauberghs M, Van de Woestijne KP. Membranous bronchioles and connective tissue network of normal and emphysematous lungs. J Appl Physiol. 1996;81:2468–80.
6. Weibel ER. Morphometry of the human lung. New York: Academic; 1963.
7. Bachofen H, Gehr P, Weibel ER. Alterations of mechanical properties and morphology in excised rabbit lungs rinsed with a detergent. J Appl Physiol. 1979;47:1002–10.
8. Gil J, Bachofen H, Gehr P, Weibel ER. The alveolar volume-to-surface ratio in air- and saline-filled lungs fixed by vascular perfusion. J Appl Physiol. 1979;47:990–1001.
9. Wilson TA, Bachofen H. A model for mechanical structure of the alveolar duct. J Appl Physiol. 1982;52:1064–70.
10. Oldmixon EH, Butler JP, Hoppin Jr FG. Dihedral angles between alveolar septa. J Appl Physiol. 1988;64:299–307.
11. Oldmixon EH, Hoppin Jr FG. Distribution of elastin and collagen in canine lung alveolar parenchyma. J Appl Physiol. 1989;67:1941–9.
12. Pattle RE. Properties, function and origin of the alveolar lining layer. Nature. 1955;175:1125–7.
13. Clements JA. Surface tension of lung extracts. Proc Soc Exp Biol Med. 1957;95:170–2.
14. Schurch S. Surface tension at low lung volumes: dependence on time and alveolar size. Respir Physiol. 1982;48:339–55.
15. Schurch S, Goerke J, Clements JA. Direct determination of surface tension in the lung. Proc Natl Acad Sci U S A. 1978;75:3417–21.
16. Schurch S, Schurch D, Curstedt T, Robertson B. Surface activity of lipid extract surfactant in relation to film area compression and collapse. J Appl Physiol. 1994;77:974–86.
17. Avery ME, Mead J. Surface properties in relation to atelectasis and hyaline membrane disease. Am J Dis Child. 1959;97:517–23.
18. Hoppin FG, Jildebrandt H. Mechanical properties of the lung. In: West JB, editor. Bioengineering aspects of the lung, vol. 3. New York: Dekker; 1977. p. 83–162.
19. Smith JC, Stamenovic D. Surface forces in lungs. I. Alveolar surface tension-lung volume relationship. J Appl Physiol. 1986;60:1341–50.
20. Bachofen H. Lung tissue resistance and pulmonary hysteresis. J Appl Physiol. 1968;24:296–301.
21. Bachofen H, Hildebrandt J. Area analysis of pressure-volume hysteresis in mammalian lungs. J Appl Physiol. 1971;30:493–7.
22. Horie T, Hildebrandt J. Dynamic compliance, limit cycles, and static equilibria of excised cat lungs. J Appl Physiol. 1971;31:423–30.
23. Suki B, Barabasi L, Lutchen K. Lung tissue viscoelasticity: a mathematical framework and its molecular basis. J Appl Physiol. 1994;76:2749–59.
24. Glazier JB, Hughes JMB, Maloney JE, West JB. Vertical gradient of alveolar size in lungs of dogs frozen intact. J Appl Physiol. 1967;23:694–705.
25. Hoffman EA. Effect of body orientation on regional lung expansion: a computed tomography approach. J Appl Physiol. 1985;59:468–80.
26. Hopkins SR, Henderson AC, Levin DL, Yamada K, Arai T, Buxton RB, Prisk GK. Vertical gradient in regional lung density in the supine human lung: the slinky effect. J Appl Physiol. 2006;103:240–8.
27. Millar AB, Denison DM. Vertical gradients of lung density in healthy supine men. Thorax. 1989;44:485–90.
28. Milic-Emili J, Henderson AM, Dolovich MB, Trop D, Kaneko K. Regional distribution of inspired gas in the lung. J Appl Physiol. 1966;21:749–59.
29. Wiener-Kronish JP, Gropper MA, Lai-Fook SJ. Pleural liquid pressure in dogs using a rib capsule. J Appl Physiol. 1985;59:597–602.

30. Lai-Fook SJ, Wilson TA, Hyatt RE, Rodarte JR. Elastic constants of inflated dog lobes. J Appl Physiol. 1976;40:508–13.
31. Liu S, Margulies SS, Wilson TA. Deformation of the dog lung in the chest wall. J Appl Physiol. 1990;68:1979–87.
32. Lai-Fook SJ, Beck KC, Southorn PA. Pleural liquid pressure measured by micropipettes in rabbits. J Appl Physiol. 1984;56:1633–9.
33. Anthonison NR, Robertson PC, Roos WRD. Gravity-dependent sequential emptying of lung regions. J Appl Physiol. 1970;28:589–95.
34. Hubmayr RD, Walters BJ, Chevalier PA, Rodarte JR, Olson LE. Topographical distribution of regional lung volume in anesthetized dogs. J Appl Physiol. 1983;54:1048–56.
35. Marcucci C, Nyhan D, Simon BA. Distribution of pulmonary ventilation using Xe-enhanced computed tomography in prone and supine dogs. J Appl Physiol. 2000;90:421–30.
36. Mead J, Takashima T, Leith DE. Stress distribution in the lungs: a model of pulmonary elasticity. J Appl Physiol. 1970;28:596–608.
37. Wilson TA. A continuum analysis of a two-dimensional mechanical model of the lung parenchyma. J Appl Physiol. 1972;33:472–8.
38. Stamenovic D. A model of foam elasticity based upon the laws of Plateau. J Colloid Interface Sci. 1991;145:255–9.
39. Stamenovic D. Micromechanical foundation of pulmonary elasticity. Physiol Rev. 1991;70:1117–34.
40. Lei M, Ghezzo H, Chen MF, Eidelman DH. Airway smooth muscle orientation in intraparenchymal airways. J Appl Physiol. 1997;82:70–7.
41. Stephens NL, Kroeger E, Mehta JA. Force-velocity characteristics of respiratory airway smooth muscle. J Appl Physiol. 1969;26:685–9.
42. Okazawa M, Pare PD, Lambert RK. Compliance of peripheral airways deduced from morphometry. J Appl Physiol. 2000;89:2373–81.
43. Gunst SJ, Stropp JQ. Pressure-volume curves and length-stress relationships in canine bronchi in vitro. J Appl Physiol. 1988;64:2522–31.
44. Lai-Fook SJ. A continuum mechanics analysis of pulmonary vascular interdependence in isolated dog lobes. J Appl Physiol. 1979;46:419–29.
45. Gunst SJ, Warner DO, Wilson TA, Hyatt RE. Parenchymal interdependence and airway response to methacholine in excised dog lobes. J Appl Physiol. 1988;65:2490–7.
46. Anafi RC, Wilson TA. Airway stability and heterogeneity in constricted lungs. J Appl Physiol. 2001;91:1185–92.
47. West JB, Matthews FL. Stresses, strains, and surface pressures in the lung caused by its weight. J Appl Physiol. 1972;32:32–345.
48. Lum H, Mitzner W. A species comparison of alveolar size and surface forces. J Appl Physiol. 1987;62:1865–71.

Chapter 2
The Chest Wall and the Respiratory Pump

Abstract The design of the respiratory pump is different from the design of the heart. The ribs carry compressive stresses that balance the pressure difference across the chest wall, and the rib cage provides a mechanism for transforming muscle shortening into chest wall expansion. The external intercostal muscles in the cranial-dorsal quadrant and the parasternal muscles have inspiratory actions and the internal intercostals in the caudal-dorsal quadrant and the triangularis sterni have expiratory actions. The diaphragm separates the pleural cavity and the abdomen, and the combined effect of diaphragm muscle tension and curvature balances the difference between abdominal pressure and pleural pressure. The inspiratory action of the diaphragm is the result of the descent of the diaphragm that accompanies muscle shortening and of the inspiratory forces that the diaphragm exerts on the lower ribs. Compartmental models provide simplified descriptions of the mechanics of the chest wall. The work of breathing is small compared to the work of the heart.

2.1 Design of the Respiratory Pump

Respiratory gasses are transported between the ambient air and the tissues of the body by a two-fluid system: gaseous transport between the atmosphere and the lungs, and transport via blood between the lungs and the tissues. Pumps are needed to drive the flows of both fluids. The heart beats five times as fast as the respiratory pump, but the stroke volume of the respiratory pump is five times that of the heart, and because the oxygen carrying capacity of air and blood per unit volume are about the same, the rates of fluid displacement for the two pumps are about the same. However, the driving pressure developed by the heart is 20 times that developed by the respiratory pump, and hence the work done is 20 times greater. This difference is ultimately due to the higher viscosity of water than air (by a factor of 25) and the much lower diffusivities of metabolic gases in water than in air (by a factor of 10^4). In fish, the relationship between the two pumps is quite different.

Although the respiratory pump is as essential as the circulatory pump, the respiratory pump has received little attention compared to the attention given to the heart. Why is that? Perhaps because the muscles of the respiratory pump are

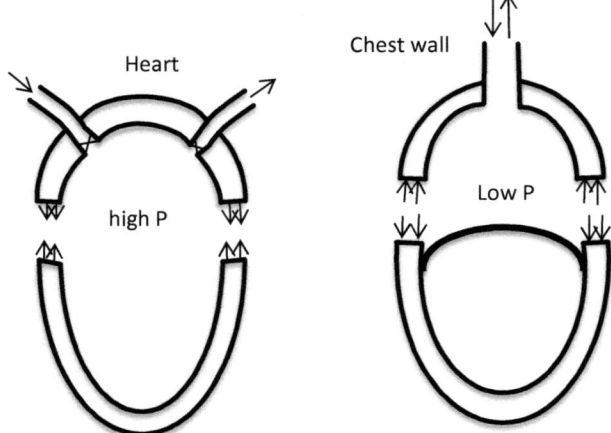

Fig. 2.1 Comparison between the designs of the circulatory and respiratory pumps

dispersed and multi-functional whereas the heart provides a clear focus. Probably, the more important reason is that the respiratory pump rarely fails, except through failure of innervation due to accidental nerve damage or degenerative nerve or muscle disease, whereas heart failure is a prominent medical problem.

The two pumps have different designs. These designs are illustrated in Fig. 2.1. The heart is a flow-through system with valves. In the heart, pressure is increased in the fluid in the chamber, and flow is driven by contraction of the volume of the chamber. Tension is required in the wall of the heart to balance the pressure in the chamber, and work is done as the heart contracts. Both of these requirements on heart wall mechanics match the capabilities of muscle; muscle develops tension when it is activated, and muscles do work when they contract. Therefore, the wall of the heart can be made of muscle.

In contrast, the respiratory pump is a batch-processing system with the same path for inflow and outflow. In the respiratory pump, pressure is reduced in the thoracic cavity to draw air into the lungs, and flow is driven by expansion of the chest wall. The wall of the chamber must therefore carry compressive stress, and the wall must expand as it does work. These requirements are contrary to the capability of muscles. This problem is solved in two ways. First, part of the chamber wall is inverted so that it is concave outward. In this part, tension in the wall balances the low pressure inside the chamber and chamber volume expands as the wall contracts. Thus, this part, the diaphragm, can be made of muscle and tendon. However, part of the wall must be concave inward. If this part is to participate in the pumping action, bones are required to carry the compressive stress, and a mechanism is required to convert muscle contraction into volume expansion [1]. The ribs and sternum fulfill these requirements.

2.1 Design of the Respiratory Pump

The chest wall consists of the structures that bound the lungs and participate in breathing: the rib cage, heart, diaphragm, and abdomen. To be compatible with the variables that are used to describe the mechanics of the lung, pressure and volume are used to describe the mechanics of the chest wall. The passive chest wall and the pleural cavity expand as pleural pressure on the inner surface of the chest wall is increased. The pressure—volume curve for the passive chest wall (CW) is shown in Fig. 2.2 along with pressure-volume curves for the lung. In this figure, lung volume (V_L), expressed as a fraction of total lung capacity (TLC), is plotted against pleural pressure (P_{pl}). The lung and chest wall must have the same volume and be exposed to the same value of P_{pl}, and the equilibrium state for the system is given by the intersection between the $P_{pl} - V_L$ curves for the lung and chest wall. Pleural pressure at this equilibrium point is approximately −4 cm H$_2$O, and lung volume, denoted functional residual capacity (FRC), is about 40 % TLC.

During passive inflation of the respiratory system, pressure at the airway opening (P_{ao}) is increased. This shifts the pressure volume curve of the lung to the right and raises the equilibrium values of P_{pl} and V_L. During active breathing, the chest wall is expanded or compressed by action of the respiratory muscles. To describe the action of the muscles in Fig. 2.2, their action must be described as effective inspiratory or expiratory pressures exerted by the muscles, denoted P^m_{insp} or P^m_{exp}. P^m_{insp} shifts the curve for the chest wall to the left, raises V_L, and lowers P_{pl}. P^m_{exp} shifts the curve to the right. Much of the study of the respiratory pump focuses on evaluating the effective pressures exerted by the respiratory muscles.

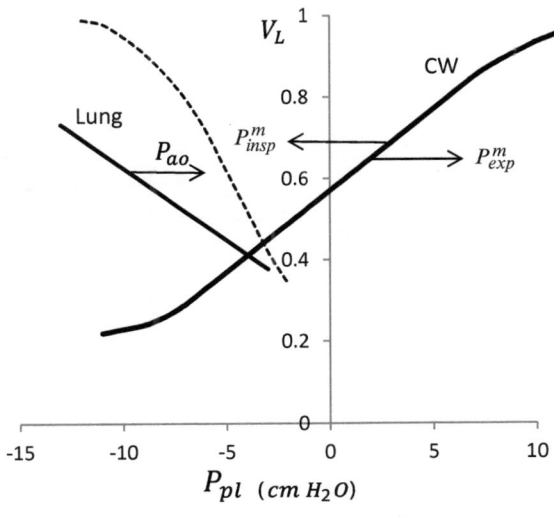

Fig. 2.2 Lung volume (V_L) as a fraction of TLC vs. pleural pressure (P_{pl}) for the passive chest wall (CW) and the lung for tidal volume oscillations (*solid line*) and deflation from TLC (*dashed line*)

2.2 Rib Cage and Intercostal Muscles

The head of each rib inserts into facets on the edges of two adjacent vertebrae. The neck of the rib extends from the head to a connection between matching facets on the rib and on the transverse process of the lower vertebra and then on to the posterior angle of the rib. Thus, the seat at the head of the rib and the connection to the transverse process define an axis around which the rib rotates. However, the facets at the transverse process are shallow and may allow some translational as well as rotational displacements. The shaft of the rib forms a smooth arc that extends from the posterior angle to the connection to the lateral end of the costal cartilage. The arc of the shaft lies nicely in a plane over most of its length with some torsion near the ventral end. In dogs, the widths of the ribs and interspaces are ~0.4 and 1.3 cm; in humans, 1.2 and 2.3 cm. The costal cartilage links the ends of the ribs to the sternum. In dogs, the costal cartilage slants sharply caudally from the sternum to the rib. In humans, it is nearly horizontal for the upper ribs and angles downward with increasing rib number. In humans, the cartilage connections for ribs 7–10 do not extend to the sternum; they connect to the cartilage of the rib above, and the ends of ribs 11 and 12 are unconnected.

The system for describing the geometry of the ribs is shown in Fig. 2.3, and the values of the parameters for the rib cage of 12 kg dog [2] and human ribs [3] are given in the Appendix. Center lines of the ribs at FRC and TLC are shown in Fig. 2.4, and the values of α and β at FRC and the rotations $\delta\alpha$ and $\delta\beta$ from FRC to TLC are shown in Fig. 2.5. These are denoted the "pump handle" and "bucket handle" rotations, respectively.

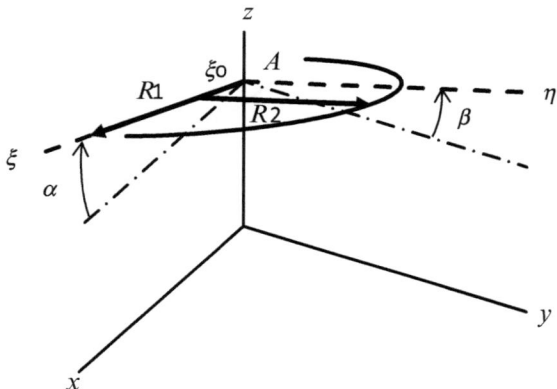

Fig. 2.3 Parameters used to describe the geometry of the ribs on the left side of the rib cage. The x axis runs ventrally and the y axis laterally. The ribs lie in the ξ–η plane which intersects the z axis at A. The ξ axis is inclined to the x axis by angle α and the η axis is inclined to the y axis by angle β. In dogs, β is negative and in humans α is negative. The line of the rib in the plane is given by $\xi = \xi_0 + R_1 \cdot \sin\theta$, $\eta = R_2 \cdot \cos\theta$

2.2 Rib Cage and Intercostal Muscles

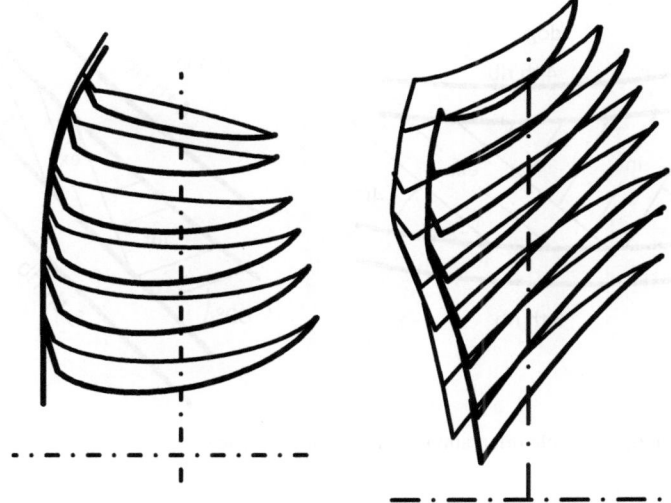

Fig. 2.4 Lateral view of the center lines of ribs 3–8 of the dog (*left panel*) and ribs 2–9 of human (*right panel*) at FRC (*heavier lines*) and TLC (*lighter lines*)

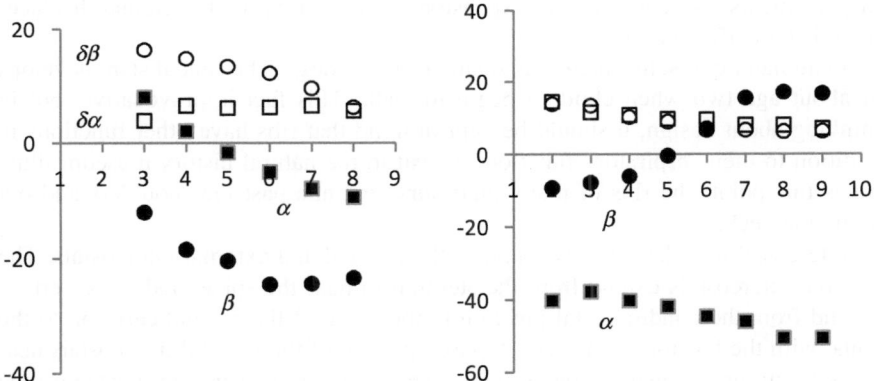

Fig. 2.5 Values of α (*filled squares*) and β (*filled circles*) at FRC and changes in the angles (*open symbols*) for rib rotations from FRC to TLC vs. rib number in the dog (*left panel*) and human (*right panel*)

It can be seen from these figures that in the dog, the ribs slant laterally and in humans, ventrally. In the dog, $\delta\beta$ is larger than $\delta\alpha$, and in humans, they are about equal. The magnitudes of total rib rotation are about the same in both species, decreasing from ~17° for the third rib to ~10° for the eighth.

The only component of rib displacement that acts to increase the volume of the thorax is the component normal to the surface. In dogs, the lateral surface area is bigger than the ventral area, the lateral slant is greater, and the lateral displacements

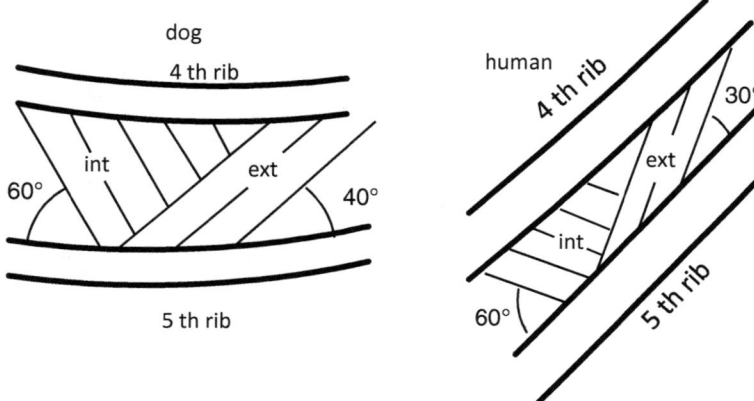

Fig. 2.6 Intercostal muscle orientation in the fourth interspace

are larger than the ventral displacements. As a result, the rib cage expands primarily through expansion in the lateral direction. In humans, the ventral surface is larger than the lateral surface, the ventral slant is larger, and the thorax expands primarily in the dorso-ventral direction. Because the rib cage tapers inward cranially, cranial displacements also contribute to expansion of the thorax, but the cranial displacement is less effective [4].

In human infants, the angle α is small, as it is in dogs. The caudal slant develops at about age two when children begin to walk. This fact is provocative, but in thinking about design, it should be kept in mind that ribs have other functions in addition to their respiratory function. A visit to the natural history museum illustrates this point; the ribs of many dinosaurs continue past the shoulders and out along the neck.

Intercostal muscles lie in two layers, the internal and external intercostals. The internal intercostals extend from the sternum to near the spine, and the externals extend from the condro-costal junction to the angle of the rib and carry on to the spine with the levator costae. The exposed portion of the internal intercostals near the sternum are termed the parasternals. The geometry of ribs and muscle on the lateral surface of the fourth interspace in the two species are shown in Fig 2.6. In both species the lines of the muscle bundles of the two layers are nearly orthogonal.

2.2.1 Respiratory Effect of the Muscles

The respiratory action of the intercostals has been a conundrum. Galen thought that both layers have an expiratory effect. Leonardo argued that the externals would raise the ribs and have an inspiratory effect, and that the internals would have an expiratory effect. Subsequently, the remaining two possible functional

2.2 Rib Cage and Intercostal Muscles

combinations were proposed. It has not been possible to activate individual segments of the individual layers of the intercostal muscles to experimentally determine their respiratory effect.

Recently, these questions have been resolved, and new data have been obtained that improve our understanding of intercostal action. A theorem of mechanics that governs linear elastic systems, applied to the respiratory system, yields the following equations [5].

$$\Delta Pao = m \cdot \sigma \cdot \mu \qquad \mu = [dl/l \cdot dVL]_{rel} \tag{2.1}$$

The first of these equations is simply a definition of the mechanical advantage of a muscle, denoted μ. Thus, μ is defined as the change in Pao per unit active stress σ and per unit muscle mass m when the muscle is activated with the airway occluded. The second equation provides the method for evaluating μ. This equation states that μ equals the fractional change in muscle length, dl/l, per unit change in lung volume, dV_L, during passive inflation of the lungs (rel). For muscle mass expressed in g, σ in kg/cm^2, and μ in L^{-1}, Eq. (2.1) yields ΔPao in cm H$_2$O.

Equation (2.1) provides a method for assessing the respiratory effect of the respiratory muscles. Qualitatively, muscles that shorten during passive inflation have negative values of μ and have an inspiratory effect, and muscles that lengthen have an expiratory effect. The quantitative maximum potential respiratory effect is obtained by measuring muscle mass and fractional change in length per unit volume increase during passive inflation and multiplying the product $m \cdot \mu$ by the maximum active stress in skeletal muscles, 3.0 kg/cm^2. For muscles that can be individually activated in the dog, namely, the parasternals in different interspaces, the scalenes, sternomastoids, and triangularis sterni, the measured ΔPao agrees extremely well with that predicted from Eq. (2.1) [6].

The rib displacements described above produce shear strains in the interspaces and compression of the upper interspaces and dilatation of the lower interspaces. The fractional length change of a muscle depends on interspace number, position around the circumference of the rib, and muscle orientation. μ was evaluated in dogs by inserting screws at the ends of muscle bundles and directly measuring the distance between the screws at FRC and after increasing lung volume to TLC [7, 8]. In humans, μ was evaluated by calculating muscle lengths from the data on rib geometry at FRC and TLC [3, 9]. The results for dogs are more certain and more detailed and the level lines for the distribution of μ for the intercostal muscles of dogs are shown in Fig. 2.7. The values in Fig. 2.7 are average values over the range of lung volumes from FRC to TLC.

The external intercostals have a potential inspiratory effect in the cranial-dorsal region of the rib cage and a potential expiratory effect in the caudal-ventral region. The internals have a potential expiratory effect in the caudal-dorsal region, and the parasternals have a strong potential inspiratory effect. Thus, both the externals and internals have potential effects of both signs, and their functional action depends on the distribution of muscle activation during inspiration and expiration.

Fig. 2.7 Level lines for the distribution of mechanical advantage (μ) of the external and internal intercostal muscles of the dog. The data described here are for 24 kg dogs

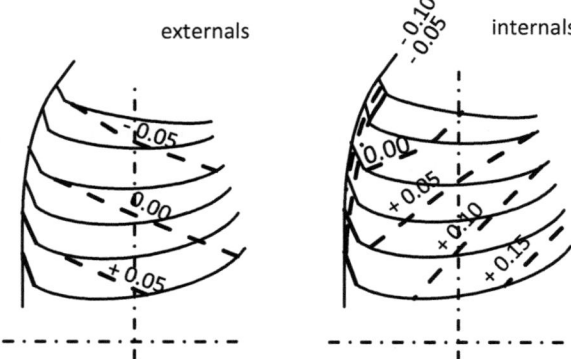

Only the external intercostals with inspiratory effect and the parasternals are activated during inspiration and only the internal intercostals with expiratory effect are activated during active expiration. Moreover, the distribution of activation mirrors the magnitude of mechanical advantage [10–12]. In the external intercostals, the level of inspiratory activation is greatest in the cranial-dorsal corner, and in the parasternals, in the upper interspaces and closest to the sternum. With increasing inspiratory effort, the level of activation in those regions increases and activation spreads to regions with a smaller magnitude of inspiratory mechanical advantage. Furthermore, the distribution of muscle mass mimics the distribution of mechanical advantage. The thickness of the externals is greatest in the cranial-dorsal corner and decreases ventrally and caudally. The thickness of the parasternals is greater than that of the external intercostals and decreases laterally and caudally. The calculated maximum total inspiratory effect of the intercostals is -18 cm H_2O. This agrees with the observation of DiMarco et al. [13, 14] for the value of ΔPao during maximal activation of the inspiratory intercostals. This maximum effect decreases with increasing lung volume.

The distribution of mechanical advantage of the human intercostals is like that of the dog. However, whereas the parasternals have a stronger potential effect than the externals in the dog, the opposite is true for humans. The magnitudes of the mechanical advantages of humans are lower than those of the dog by a factor of ~3, and the masses are larger by the same factor; the maximum total effect is about the same.

2.2.2 Mechanisms of Intercostal Muscle Action

Hamberger's theory of intercostal muscle action was presented in his book, published in 1734 [15]. The heart of the book lies in diagrams like that shown in Fig. 2.8. The spine is represented by the vertical bar on the right and two ribs by bars

2.2 Rib Cage and Intercostal Muscles

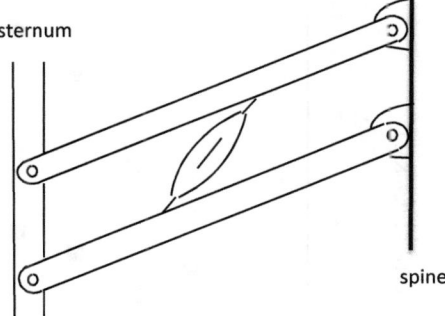

Fig. 2.8 Hamberger model for the mechanics of intercostal muscle action

connected to the spine by pin joints and slanting down to connections to the sternum. An intercostal muscle slants between the two ribs. The normal component of the force exerted on the lower rib exerts a moment around the pin joint on the spine, and the normal component of the equal and opposite force exerted on the upper rib exerts a moment with opposite sign on that rib. Because the moment arm of the force on the lower rib is greater than that on the upper rib, the net moment would cause a cranial rotation of the ribs. This mechanism will be referred to as the Hamberger mechanism. By this analysis, muscles that slant upward to the right of the vertical in Fig. 2.8, as the external intercostals do, would cause a cranial displacement of the ribs and act as inspiratory muscles, and muscles that slant to the left, the internal intercostals, would have an expiratory effect. The conclusion that follows from this model is that all external intercostals have an inspiratory effect and all internal intercostals have an expiratory effect.

Although alternate ideas about intercostal muscle action were expressed throughout the intervening years, the Hamberger model has provided the dominant concept of intercostal mechanics for more than two centuries. The Hamberger model in fact correctly captures one of the mechanisms of intercostal muscle action, but the model has shortcomings. First, the model is two-dimensional whereas the rib cage is three-dimensional. In the diagram, the plane of the ribs is perpendicular to the axis of rib rotation. In the rib cage, the plane of the ribs on the dorsal surface of the rib cage is perpendicular to the axis of rotation, but the angle changes moving ventrally around the rib. On the ventral surface, the plane is parallel to the axis, and the difference in moments on the two ribs is zero. Near the sternum, the net moment changes sign and becomes expiratory. This mechanism explains, in part, the inspiratory effect of the external intercostals and expiratory effect of the internal intercostals on the dorsal surface and the dorsal-ventral gradients in respiratory effect within each interspace.

A second shortcoming of the Hamberger model is that it ignores the coupling between rib displacement and lung expansion. A second mechanism was revealed when the coupling between ribs and lung in dogs was studied by De Troyer et al. [16–18]. Screws were inserted into the ribs on both sides of the rib cage, a yoke was attached to the two screws in the nth rib pair, and airway opening pressure

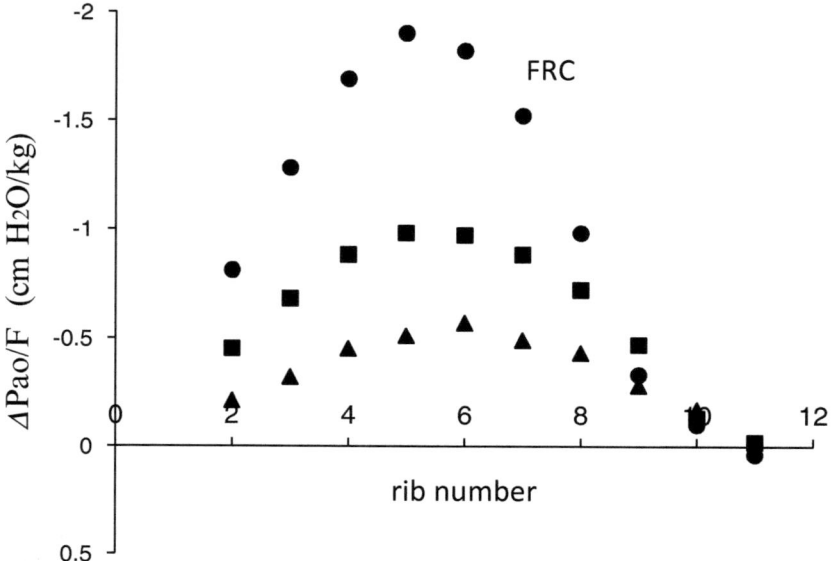

Fig. 2.9 Values of ΔPao/F for external forces (F) applied to rib pairs at FRC (*circles*) and higher lung volumes (*squares* and *triangles*)

was measured as a cranial force was applied to the rib pair with the airway occluded. The results of this experiment are shown in Fig. 2.9. It can be seen that the respiratory effect per unit force varies markedly with rib number. It increases with rib number from the second to fifth rib pair and then decreases to essentially zero for the 11th pair. Rib displacements were also measured, and these were nearly uniform, increasing slightly with increasing rib number. Thus, the strong dependence of Δ*Pao* on *n* is not the result of differences in rib compliance. Presumably, it is the result of the dependence of the lung-apposed surface area subtended by the ribs on rib number. For the upper ribs, rib radius and circumference increase with rib number, and thus, the subtended surface area increases with rib number. The lung-apposed area then decreases for the more caudal ribs because the fraction of the subtended area that is apposed to the lung decreases and the fraction that is apposed to the abdomen increases. At higher lung volumes, the dependence of *Pao/F* on rib number is similar, but the magnitudes decrease markedly with increasing lung volume.

The Hamberger mechanism for muscle action depends on the slant of the intercostals and the resulting difference between the moments exerted on the two ribs that bound the interspace. Figure 2.9 shows that a muscle that is oriented axially would also have a net respiratory effect because the respiratory effects of equal and opposite forces on adjacent ribs are not equal. In the cranial interspaces, the inspiratory effect on the lower rib is greater than the expiratory effect on the upper rib, and the net effect is inspiratory. In the caudal interspaces, the opposite holds. Thus, this mechanism accounts for the cranial-caudal gradient in respiratory effect from generally inspiratory in cranial interspaces to generally expiratory in the caudal interspaces.

2.3 Diaphragm

The diaphragm is a thin membrane that separates the thoracic cavity from the abdomen. It consists of the central tendon, a thin sheet of connective tissue that is functionally inextensible, surrounded by a muscle sheet. The muscle bundles that form this sheet originate at the edge of the central tendon and insert on the upper edge of the lower ribs and on the spine as shown in Fig. 2.10.

In the dog, the central tendon is roughly triangular with the apex at the ventral end and the curved base at the dorsal end. Lines of connective tissue extend from the corners of the base to the spine. The central tendon covers the dome of the diaphragm that is located in the mid-plane. The muscle sheet consists of two parts, the costal diaphragm that extends from the sternum to the most dorsal points on the arcs of the ribs and the crural diaphragm that extends from the dorsal points to the midplane. The muscles bundles of the costal diaphragm insert on the upper edge of the lower ribs, and the bundles of the crural diaphragm insert on the spine. In a number of studies, the length of the diaphragm muscle bundles and the height of the zone of apposition have been underestimated because a band of fatty tissue on the abdominal surface of the diaphragm was identified as the line of insertion. Although the muscles are not visible from the abdominal side below this band, they extend 1–2 cm beyond the fatty band to the true line of insertion [19]. In a 24 kg dog, the length of the muscle bundles is ~4 cm at the ventral end, increasing to ~8 cm on the lateral surface [20]. The central tendon and the muscle sheet near the tendon are apposed to the lung. The distal part of the muscle sheet lies on the inner surface of the rib cage and this region is called the zone of apposition. The line of the top of the zone of apposition (the bottom edge of the lung) lies in a plane that slants downward

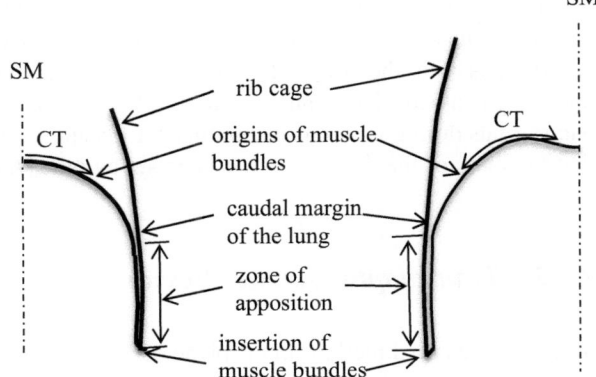

Fig. 2.10 Transverse sections through the domes of the diaphragm in the dog (*left*) and human (*right*). The sagittal midplanes (SM) are shown by *interupted lines* and the central tendons (CT) by *arrows*

from the transverse plane by ~30°. With increasing lung volume, the muscle shortens, the dome of the diaphragm descends, the rib cage expands and the height of the zone of apposition decreases.

In humans, the central tendon is two-lobed with one lobe covering the dome in each hemi-diaphragm and some fibers of connective tissue connecting the lobes across the saddle point in the mid-plane. The length of the muscle bundles is ~6 cm at the ventral end and increases rapidly to ~13 cm at the lateral surface. MRI data for humans [21] provide detailed information about the shape of the human diaphragm and the displacements that occur with lung inflation. With inflation from FRC to TLC, the dome descends by ~5 cm and the muscles shorten by ~35 %. The height of the zone of apposition on the lateral surface decreases from ~9 cm at FRC to 4.5 cm at TLC. At FRC, seventy percentage of the muscle lies in the zone of apposition, and this decreases to 50 % at TLC.

Mead [22] ascribed four functions to the zone of apposition. First, it provides a reservoir of rib cage surface area so that the base of the lung can descend, relative to the rib cage, without distorting the shape of the lung. Second, expansion of the rib cage in the area apposed to the diaphragm provides a pathway, in addition to the expansion of the abdominal wall, for the volume displaced by the diaphragm. Third, it serves as the site for the application of abdominal pressure to the lower rib cage. Fourth, it serves to direct the force exerted by the diaphragm at the line of insertion in the cranial direction. A fifth function could be added to this list; it provides a reservoir of muscle length so that, for a given descent of the diaphragm, fractional muscle shortening is reduced.

2.3.1 Respiratory Effect

The muscles of the diaphragm shorten by 30–35 % with passive lung inflation from FRC to TLC [19, 21, 23]. For a 24 kg dog, therefore, $\mu = -0.25 L^{-1}$. Muscle mass is 95 g [24], and the maximum respiratory effect of the diaphragm is $P_{ao} = -70$ cm H_2O. The proportionality between level of activation and mechanical advantage, described above, applies within each muscle group, but not across groups. In the dog, the mechanical advantage of the diaphragm is twice that of the parasternals, but the level of activation of the diaphragm during spontaneous breathing is always less than maximum [25] whereas the level of activation of the parasternals during vigorous inspiratory efforts approaches 100 %. Thus, it appears that the mass of the diaphragm has been set by other functions.

2.3.2 Transdiaphragmatic Pressure

Strains in the two directions in the plane of the muscles of the diaphragm have been measured during spontaneous breathing [26]. Whereas the muscle layer shortens along the line of the muscle bundles, the strain in the direction transverse to the line

2.3 Diaphragm

of the muscle is essentially zero. The layer is compliant in the transverse direction [27], and it follows that, *in vivo*, the stress in the transverse direction is zero. Thus, the muscle layer is a membrane that carries stress in only one direction, the direction of the muscle fibers. It follows from this that the lines of the muscles must lie along geodesics of the surface. That is, the curvature of the lines is perpendicular to the surface. The lung-apposed diaphragm can be roughly pictured as a slinky laid out around the inner surface of the rib cage with the opening within the ring covered by the central tendon and the tangent line between the slinky and the rib cage forming the upper edge of the zone of apposition [28].

It also follows that the equation of equilibrium for the lung-apposed muscle layer is the following where P_{di} is transdiaphragmatic pressure, t is the thickness of the layer, σ is the stress in the muscle, and ρ is the radius of curvature of the line of the muscle bundle.

$$P_{di} = t \cdot \sigma / \rho \tag{2.2}$$

For the dog, $t = 0.17$ cm [29], $\rho = 4.5$ cm [30], and for maximum stress, 3 kg/cm^2, $P_{di} = 115$ cm H$_2$O. This agrees with the value reported by Road et al. [31] for P_{di} for maximum activation of the diaphragm at FRC. The value of ρ remains constant as lung volume increases and muscle length decreases down to 60 % Lo [32], and the measured curve of P_{di} vs. muscle length corresponds to the curve of σ vs. L/Lo for skeletal muscle.

The value of P_{di}, calculated here, is not the same as the value of P_{ao} for the diaphragm calculated earlier. The value of P_{di}, calculated here, is transdiaphragmatic pressure, whereas the value of P_{ao} is the airway pressure that results from activation of the muscle. The relation between the two will be described below.

The inspiratory action of the diaphragm is the result of two mechanisms. First, P_{di} is applied to the boundary between the lung and the abdomen. With increasing P_{di}, the muscles of the diaphragm shorten and the dome of the diaphragm descends, thereby expanding the volume of the thoracic cavity and the lung. With diaphragm descent, pleural pressure falls, abdominal pressure increases, and the abdominal wall expands.

The second mechanism is the result of the two forces exerted by the diaphragm on the rib cage. The first of these, the insertional force, is the axial force exerted by muscle tension acting directly at the line of insertion on the lower margin of the rib cage. This force has been recognized since Galen. The second, the appositional force, was described by Mead in 1979 [22]. This is the lateral pressure that acts on the rib cage in the zone of apposition. In the zone of apposition, the diaphragm is constrained to lie along the surface of the rib cage, and its curvature is less than in the lung-apposed region. The pressure exerted on the rib cage can be described in either of two ways: either as the pressure exerted by the diaphragm on the rib cage because the curvature of the diaphragm is reduced, or alternatively, as the partial or complete transmission of abdominal pressure across the diaphragm because the curvature of the diaphragm is less than that in the lung-opposed region.

Most experimental studies of the action of the diaphragm in dogs have used maximum activation of the phrenic nerves to stimulate diaphragm contraction. This method provides a standard reproducible level of diaphragm activation, but the results are skewed by the fact that maximum diaphragm activation is unphysiological and the distortions of the rib cage that it generates are unphysiological. In 2011, De Troyer introduced a new preparation [33]. He severed the inspiratory intercostal muscles in all interspaces, except the parasternals in one interspace, which were used to monitor neural drive, and he measured the displacements of ribs in the upper and lower rib cage during spontaneous breathing with the diaphragm alone. In this preparation, the rib cage does not move as a unit; at FRC, the upper ribs move caudally and inward and the lower ribs move cranially and outward. This is consistent with the earlier observation that the upper ribs move paradoxically in quadriplegic subjects breathing with their diaphragm alone [34, 35].

The diaphragm is frequently described as a piston that moves caudally as the muscles contract. This analogy is incomplete. The diaphragm is more like a hemi-balloon attached to the lower margin of the rib cage and in contact with the chest wall in the zone of apposition. With activation, the dome descends, and the diaphragm pulls cranially and pushes laterally on the lower rib cage, thereby acting as the primary inspiratory muscle for the lower rib cage.

2.3.3 Volume Dependence

The respiratory effect of both the intercostal muscles [13] and the diaphragm [36] decrease markedly with increasing lung volume. The decrease of intercostal effect is primarily due to the change in the geometry of the rib cage [37]; for increased β in the dog or α in man, the component of rib displacement perpendicular to the surface decreases for a given increment in β or α. The decrease in diaphragm effect is due to the decrease in muscle length, and hence muscle force, for a given level of activation.

2.4 Other Respiratory Muscles

In addition to the external intercostals in the cranial interspaces and the parasternals, the pectoralis major and pectoralis minor that lie on the ventral surface of the upper rib cage are activated during inspiration, but only during extreme efforts.

Two of the muscles of the neck act as inspiratory muscles, the scalenes and the sternomastoids. In humans, the scalenes run from the transverse processes of the lower cervical vertebrae to the upper surface of the first two ribs. The sternomastoids run from the mastoid process to the manubrium sterni and the clavicle. In dogs, both muscles are silent during quiet breathing and are recruited during more forceful inspiration. In humans, the scalenes are active during quiet breathing and the sternomastoids are recruited during more forceful inspiration. These muscles add a potential maximum inspiratory effect of ≈ -6 cm H_2O [38, 39].

In addition to the internal intercostals in the caudal interspaces, the triangularis sterni acts as an expiratory muscle of the rib cage. This muscle lies on the inner surface of the ventral rib cage. It extends from the dorsal part of the sternum to the costal cartilages of the third to seventh ribs. It is activated during all expiratory efforts.

The muscles of the abdominal wall consist of four layers: from outermost to innermost, the rectus abdominus, the external oblique, the internal oblique, and transversus abdominus. Of these, the transverse abdominus and the internal oblique have large expiratory mechanical advantages [40]. They exert an expiratory pressure on the abdominal wall and a caudal force on the lower ribs, thereby raising abdominal pressure and driving the dome of the diaphragm up into the thoracic cavity.

During normal breathing, only inspiratory muscles are activated during the inspiratory phase and only expiratory muscles during the expiratory phase. However, during phonation, and especially trained singing, both inspiratory and expiratory muscles are activated, presumably to increase the control of the pressure that drives the flow that drives vocal cord vibrations and the production of sound.

2.5 Compartmental Models

The total number of degrees of freedom of the chest wall is large [41], and the number of elastic parameters that would be required in order to construct a detailed quantitative model of the chest wall is very large. However, the distribution of activation of each group of respiratory muscles is fixed by the neurological controls, and thus the range of potential displacements is limited. Portions of the chest wall have been conceptually combined into compartments to obtain manageable models. These models provide an expression of our understanding of the mechanics of the chest wall and a vehicle for expressing quantitative relations between different observations.

In the one-compartment model, the chest wall is pictured as a structure with a configuration that is described by one degree of freedom, thoracic cavity volume. This model is depicted in the left panel of Fig. 2.11, in which the chest wall is a piston in a cylinder in series with a second piston that represents the lung, separated by the incompressible pleural space.

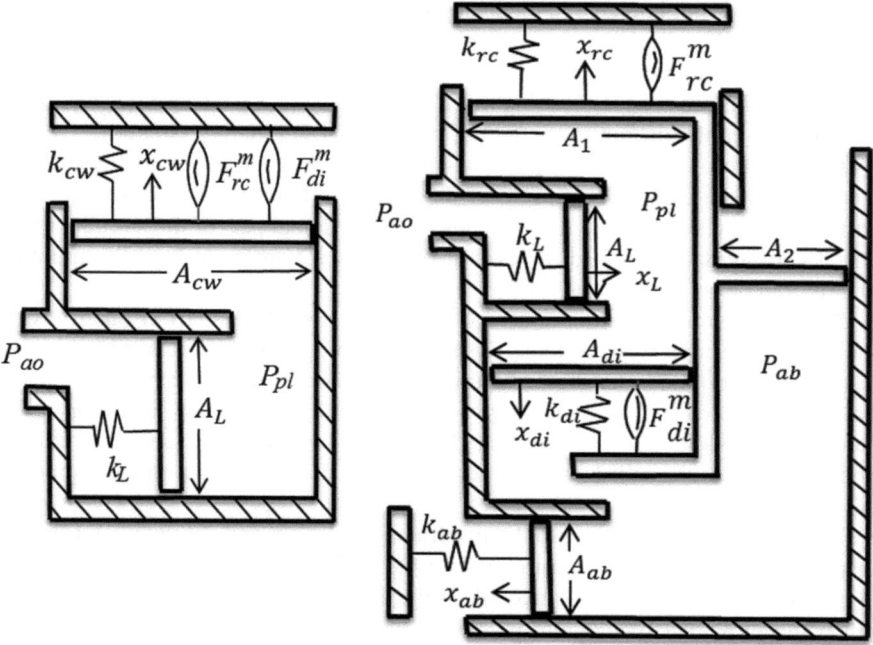

Fig. 2.11 Schematic diagram of the one-compartment model (*left panel*) and two-compartment model (*right panel*) of the chest wall. The hatched bars depict fixed walls, and the open bars depict pistons that represent the rib cage (*rc*), lung (*L*), diaphragm (*di*) and abdomen (*ab*). The pistons with area A_i are connected to the foundation with springs with spring constants k_i. The displacements of the pistons, x_i, are driven by active forces in the muscles of the rib cage (F_{rc}^m) and diaphragm (F_{di}^m) and by pressure applied at the airway opening (P_{ao}). All displacements and pressures are differences from positions and pressures at FRC

In the one-compartment model, for muscle activation with the airway closed, $P_{ao} = P_{pl} = -P^m$. From the comparison of this equation with Eq. (2.1), it follows that for this model, $P^m = -m \cdot \sigma \cdot \mu$ for both the muscles of the rib cage and the diaphragm.

In 1967, Konno and Mead [42] noted that one could inspire either by primarily expanding the rib cage or by primarily displacing volume with the abdominal wall, and at a fixed lung volume, one could shift volumes between these two. They measured the ventral displacements of a number of points on the ventral surface of the rib cage and abdomen for different breathing maneuvers. The relative displacements of points on the rib cage were nearly the same for all maneuvers; less so for the abdomen. By combining data for passive inflation and iso-volume maneuvers, they calibrated the displacements of one point on the rib cage and one on the abdomen so that the displacements of these points served as volume transducers. Since this work, the concept of the chest wall as a two-compartment system has been a cornerstone of the description of the mechanics of the chest wall.

2.5 Compartmental Models

A schematic diagram of the two-compartment model is shown in the right panel of Fig. 2.11. The insertional force exerted by the diaphragm on the rib cage is represented by the force acting on the shelf below the diaphragm. The appositional force is represented by abdominal pressure acting on area A_2. The equilibrium equations for the four pistons in this model are the following.

$$P_{pl} + P_{rc}^m + \alpha \cdot P_{di} = E_{rc} \cdot V_{rc} \qquad (2.3)$$

$$P_{di}^m - P_{di} = E_{di} \cdot (V_{ab} + \alpha \cdot V_{rc}) \qquad (2.4)$$

$$P_{pl} + P_{di} = E_{ab} \cdot V_{ab} \qquad (2.5)$$

$$P_{ao} - P_{pl} = E_L \cdot V_L \quad \text{where} \quad V_L = V_{rc} + V_{ab} \qquad (2.6)$$

The parameters in these equations are related to those shown in Fig. 2.11 by the following equations: $P_i^m = F_i^m/A_i$, $E_i = k_i/A_i^2$, $V_i = A_i \cdot x_i$, $A_{rc} = A_1 + A_2$, $\alpha = (A_{di} + A_2)/A_{rc}$, and P_{di} denotes transdiaphragmatic pressure, $P_{ab} - P_{pl}$. The volumes in the nether region are related by the equation, $V_{di} = V_{ab} + (A_2/A_{rc}) \cdot V_{rc}$.

Equation (2.3) is the same as the equation of equilibrium for the rib cage given by Loring and Mead [43]. In this equation, the parameter α describes the effective pressure exerted on the rib cage by the diaphragm as a fraction of P_{di}. Loring and Mead report the results of experiments intended to measure the value of α. In these experiments, V_{rc}, P_{pl}, and P_{ab}, were measured for passive inflation and during a maneuver in which the subjects inspired predominantly with their diaphragm. The average value of α at FRC was ≈ 0.6. They note that subjects did not totally suppress intercostal activation during inspiration, and their value of α is presumably higher than the true value. Here, we take $A_2/A_{rc} = 0.3$, $A_{di}/A_{rc} = 0.2$, and $\alpha = 0.5$.

To represent the properties of a 24 kg dog with an inspiratory capacity of 1.2 L, the values of the elastances are taken as the following, all in cm H_2O/L.

$$E_{rc} = 10 \quad E_{di} = 8 \quad E_{ab} = 18$$

These values were chosen so that the predictions of the model for passive inflation of the chest wall match three observations: $V_{di} = 0.5 \cdot \Delta V_L$ [44], $P_{ab}/P_{pl} = 0.5$ [45], and $E_{cw} = 10$ cm H_2O/L. For a 72 kg human with an inspiratory capacity of 3.6 L, the values of the E's are 1/3 those given for the dog.

Coordinated breathing with the intercostals and diaphragm is modeled by applying both P_{rc}^m and P_{di}^m with the airway open ($P_{ao} = 0$). Agostoni et al. [46] pointed out that the configuration of the chest wall during passive inflation is the configuration for which the elastic energy stored in the chest wall is minimum for a given chest wall volume. Therefore, the trajectory for passive inflation is the trajectory for which the work of chest wall expansion is minimum. It can be seen from Eqs. (2.3–2.6) that the chest wall is driven along the relaxation trajectory for $P_{rc}^m = (1 - \alpha) \cdot P_{di}^m = 0.5 \cdot P_{di}^m$. For these values of P^m and $E_L = 12$ cm H_2O/L, the model predicts $P_{ab}/P_{pl} = -0.4$, in agreement with observations [47].

For the case of muscle activation with the airway closed, the numerical solution to Eqs. (2.3–2.6), with the parameter values listed above, is the following.

$$P_{ao} = -0.73 \cdot P_{rc}^m - 0.63 \cdot P_{di}^m \tag{2.7}$$

Equation (2.7) has the form, $P_{ao} = -k_1 \cdot P_{rc}^m - k_2 \cdot P_{di}^m$. By comparison between this equation and Eq. (2.1), it can be seen that the muscle pressures in the two-compartment model are given by the following equations.

$$P_{rc}^m = -(1/k_1) \cdot \sum_{rc} m \cdot \sigma \cdot \mu = -1.4 \cdot \sum_{rc} m \cdot \sigma \cdot \mu$$
$$P_{di}^m = -(1/k_2) \cdot (m \cdot \sigma \cdot \mu)_{di} = -1.6 \cdot (m \cdot \sigma \cdot \mu)_{di}.$$

The maximum potential value of $\sum_{rc} m \cdot \sigma \cdot \mu$ estimated above is -24 cm H_2O. The corresponding value of P_{rc}^m is 34 cm H_2O. For coordinated muscle activation that produces no chest wall distortion ($P_{rc}^m = 0.5 \cdot P_{di}^m$), $P_{di}^m = 68$ cm H_2O, and $(m \cdot \sigma \cdot \mu)_{di} = -42$ cm H_2O. Thus, the maximum inspiratory pressure that can be generated with no chest wall distortion is -66 cm H_2O, and the diaphragm contributes 64 % of P_{ao}, which matches the estimate of D'Angelo and Bellemare [48]. The value of P_{di}^m, 68 cm H_2O, is well below the estimate of the maximum value of P_{di} (Eq. 2.2).

The two-compartment model and Eq. (2.7) also provide a reconciliation of the two descriptions of diaphragm action obtained above. By Eq. (2.1), the maximum inspiratory potential of the diaphragm was estimated as -70 cm H_2O. By Eq. (2.2), the maximum value of P_{di}^m was 110 cm H_2O. For $P_{ao} = -0.63 P_{di}^m$, as given by Eq. (2.7), these two results are consistent.

The two-compartment model describes the contribution of the diaphragm to expansion of the rib cage and hence, explains the larger contribution of the diaphragm than the intercostals to chest wall expansion. It also describes the contribution of the volume displaced by the zone of apposition to the volume balance for the abdomen. In the model, 45 % of the volume displaced by the diaphragm is taken by the expansion of the lower rib cage in the zone of apposition, in agreement with the fraction measured by Knight et al. in dogs [49] and the estimates of Mead and Loring for humans [50]. The model also depicts the fact that the diaphragm inserts on an elastic foundation and that the displacement of the foundation, as well as the displacement of the dome of the diaphragm, determine diaphragm muscle shortening. In the model, the displacement of the lower rib cage contributes 20 % of diaphragm shortening, in agreement with the value inferred from data on the displacements of the diaphragm dome and the lower ribs [17].

For coordinated breathing, the two-compartment model provides an accurate description of chest wall kinematics. For more extreme maneuvers, this model is inadequate. In subjects with tetraplegia [34, 35] and in dogs breathing with the diaphragm alone [33], the upper ribs move caudally and inward during inspiration whereas the lower ribs move cranially and outward. For these maneuvers, the chest wall has been modeled by a three-compartment model in which the rib cage is subdivided into upper and lower compartments [51, 52].

One of the forces exerted on the lower rib cage is the appositional force, and this force is proportional to the area of apposition. With increasing lung volume, the area of the zone of apposition decreases, the force exerted by the diaphragm on the lower rib cage decreases, and the lower rib cage moves inward during diaphragm activation [36]. This decrease in the effect of the diaphragm also is manifested in Hoover's sign, an inward displacement of the lower rib cage during inspiration that is seen in some patients with COPD.

These models describe the actions of the inspiratory muscles. The data that would be required to formulate comparative models for the action of the expiratory muscles have not been obtained.

2.6 Work of Breathing

The rate of work ($\dot{W}$) of the respiratory muscles for quiet breathing with breathing frequency f, tidal volume V_T, and end-expiratory volume equal to FRC is given by Eq. (2.8).

$$\dot{W} = \frac{1}{2} \cdot f \cdot E_{rs} \cdot V_T^2 \qquad E_{rs} = E_L + E_{cw} \qquad (2.8)$$

The work per breath is usually depicted by the Campbell diagram shown in Fig. 2.12.

The work per breath is the shaded area in the diagram. For quiet breathing, an additional 25 % can be added to the shaded area to account for work done against the resistive load. Also, during active inspiration, chest wall distortion from the passive configuration at a scale smaller than the compartmental scale adds another 25 % to the work done by the muscles [53]. Because the work per breath is proportional to the square of the difference between end-inspiratory volume and FRC (or end-expiratory volume and FRC), the work for a given tidal volume would be minimized if half the tidal volume were taken below FRC and half above. Dogs [4] and horses use their expiratory muscles during quiet breathing and end-expiratory volume is below FRC. Humans use only the inspiratory muscles during quiet breathing, but with the onset of exercise, the expiratory muscles are recruited and end-expiratory volume is lower than FRC. The diagram also shows the added work required if end-expiratory volume is above FRC, as occurs because of dynamic hyperinflation in COPD patients. For quiet breathing, the work of breathing is quite small, $\sim 2 \cdot 10^{-2}$ W. During exercise, this can increase by a factor of 25, but in normals, the work of breathing is always a small part of total metabolism.

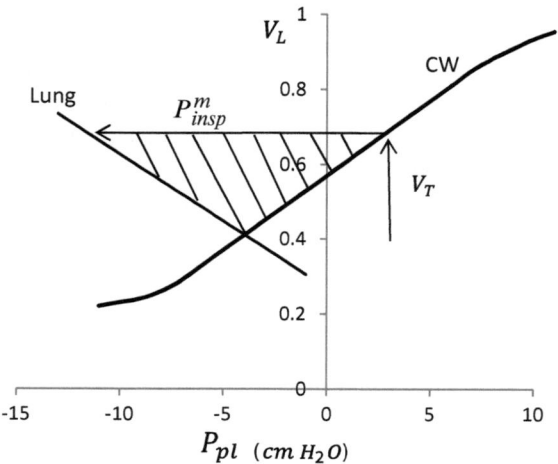

Fig. 2.12 Campbell diagram with the work done per breath shown by the *shaded area*

2.7 Mechanics of the Pleural Space

The pleural space is a thin layer of fluid that lies between the lung and the chest wall. In breathing, the ribs move cranially and the lung is stretched caudally. Thus, the lungs slide along the inner surface of the rib cage. The liquid in the pleural space lubricates this sliding. The pleural space is thin, ~ 15 μ in the dog, except along lines where a lobar fissure meets the pleural surface where a broader channel runs along the fissure. The pressure distribution in the pleural space is dictated by the mechanics of the lung and chest wall. As described in the first chapter, the pressure gradient must balance the gravitational force on the lung, $\rho_L\, g$. The magnitude of this pressure gradient at FRC is therefore about 0.25 cm H_2O/cm. This pressure gradient is too small to balance the gravitational body force $\rho\, g$ that acts on the fluid with the density of water in the pleural space, and the fluid drains downward. However, a pumping action driven by the sliding between the lung and the rib cage pumps fluid out from the channels along the lobar fissures into the thin layer on the surface of the lung [54, 55]. Fluid drains downward along the surface and flows back up along the fissure to complete the circulatory flow that maintains the pleural space. The volume of pleural fluid is controlled by absorption and emission of fluid across the parietal pleura.

Appendix

Values of the parameters of the arcs that describe the center lines of the ribs of dogs (Tables 2.1 and 2.2) and humans (Tables 2.3 and 2.4).

Table 2.1 Values of the parameters for a 12 kg dog at FRC

Rib #	A (cm)	ξ_0 (cm)	R_1 (cm)	R_2 (cm)	α	β
3	17.5	−0.2	4.7	3.6	7.9	−12.1
4	16.5	0.2	5.3	4.5	2.0	−18.5
5	14.6	0.2	5.8	5.2	−1.7	−20.5
6	13.2	0.2	6.2	5.8	−5.1	−24.5
7	11.5	0.0	6.5	6.3	−8.0	−24.4
8	9.3	−0.1	6.8	6.5	−9.5	−23.4

Table 2.2 Difference between parameter values at TLC and FRC

Rib #	δA	$\delta\xi_0$	$\delta\alpha$	$\delta\beta$
3	−0.1	0.1	3.8	16.3
4	0.0	0.0	6.4	14.7
5	0.1	0.1	6.0	13.3
6	0.2	0.0	6.1	12.1
7	0.3	0.1	6.4	9.4
8	0.6	0.0	5.5	5.5

Table 2.3 Parameter values for human ribs at FRC

Rib #	A (cm)	ξ_0 (cm)	R_1 (cm)	R_2 (cm)	α	β
2	37.8	3.6	7.2	10.6	−40.3	−23.1
3	34.4	3.1	9.2	11.6	−37.8	−21.2
4	30.5	2.5	11.7	12.7	−40.3	−16.3
5	25.9	1.5	12.8	12.6	−41.9	−9.4
6	20.3	−0.4	14.0	12.8	−44.6	−0.3
7	14.4	−2.2	14.7	13.0	−46.1	8.8
8	9.8	−2.6	15.8	13.3	−50.7	10.8
9	6.6	−3.6	14.4	13.4	−50.7	10.3

Table 2.4 Differences between values at TLC and FRC for human ribs

Rib #	δA	$\delta\xi_0$	$\delta\alpha$	$\delta\beta$
2	−1.0	−0.4	14.3	13.7
3	−0.1	−0.4	11.4	13.3
4	0.5	0.4	10.7	10.1
5	1.0	0.7	9.6	8.9
6	1.4	1.6	9.4	6.9
7	1.5	1.5	7.9	6.6
8	1.7	2.1	7.9	6.2
9	1.5	1.5	6.0	6.3

References

1. Cappello M, De Troyer A. On the respiratory function of the ribs. J Appl Physiol. 2002;92:1642–6.
2. Margulies SS, Rodarte JR, Hoffman EA. Geometry and kinematics of dog ribs. J Appl Physiol. 1989;67:707–12.
3. Wilson TA, Legrand A, Gevenois P-A, De Troyer A. Respiratory effects of the external and inter intercostal muscles in humans. J Physiol. 2001;530:319–30.
4. De Troyer A, Ninane V. Triangularis sterni: a primary muscle of breathing in the dog. J Appl Physiol. 1986;60:14–21.
5. Wilson TA, De Troyer A. Effect of respiratory muscle tension on lung volume. J Appl Physiol. 1992;73:2283–8.
6. De Troyer A, Legrand A. mechanical advantage of the canine triangularis sterni. J Appl Physiol. 1998;84:562–8.
7. De Troyer A, Legrand A, Wilson TA. Rostrocaudal gradient of mechanical advantage in the parasternal intercostal muscles of the dog. J Physiol. 1999;495:239–89.
8. De Troyer A, Legrand A, Wilson TA. Respiratory mechanical advantage of the canine external and internal intercostal muscles. J Physiol. 1999;518:283–9.
9. De Troyer A, Legrand A, Gevenois PA, Wilson TA. Mechanical advantage of the human parasternal intercostal and triangularis sterni muscles. J Physiol. 1998;513:915–25.
10. De Troyer A, Gorman RB, Gandevia SC. Distribution of inspiratory drive to the external intercostal muscles in humans. J Physiol. 2003;546:943–54.
11. De Troyer A, Legrand A. Inhomogeneous activation of the parasternal intercostals during breathing. J Appl Physiol. 1995;79:55–62.
12. Johnson Jr RL, Hsia CCW, Takeda SI, Wait JL, Glenny RW. Efficient design of the diaphragm: distribution of blood flow relative to mechanical advantage. J Appl Physiol. 2002;93:925–30.
13. DiMarco AF, Romaniuk JR, Supinski GS. Mechanical action of the interosseus intercostal muscles as a function of lung volume. Am Rev Respir Dis. 1990;142:1041–6.
14. DiMarco AF, Supinski GS, Budzinska K. Inspiratory muscle interaction in the generation of changes in airway pressure. J Appl Physiol. 1989;66:2573–8.
15. Hamberger GE. De Respirtionis Mechanismo et usu Gennino, Christoph Croeker, 1740
16. De Troyer A, Leduc D. Effects of inflation on the coupling between the ribs and the lung in dogs. J Physiol. 2004;555:481–8.
17. De Troyer A, Wilson TA. Coupling between the ribs and the lung in dogs. J Physiol. 2002;540:231–6.
18. Wilson TA, De Troyer A. The two mechanisms of intercostal muscle action on the lung. J Appl Physiol. 2004;96:483–8.
19. De Troyer A, Leduc D, Cappello M, Minne B, Rooze M, Gevenois PA, Wilson TA. Mechanisms of the inspiratory action of the diaphragm during isolated contraction. J Appl Physiol. 2009;107:1736–42.
20. Boriek AM, Rodarte JR, Margulies SS. Zone of apposition in the passive diaphragm in the dog. J Appl Physiol. 1996;81:1929–40.
21. Gauthier AP, Verbanck S, Estenne M, Segebarth C, Macklem PT, Paiva M. Three dimensional reconstruction of the in vivo human diaphragm shape at different lung volumes. J Appl Physiol. 1994;76:495–506.
22. Mead J. Functional significance of the area of apposition of diaphragm to rib cage. Am Rev Respir Dis. 1979;119S:31–2.
23. Wilson TA, Boriek A, Rodarte JR. Mechanical advantage of the canine diaphragm. J Appl Physiol. 1998;85:2284–90.
24. Leiter JC, Jacopo PM, Tenney SM. A comparative analysis of contractile characteristics of the diaphragm and of respiratory system mechanics. Respir Physiol. 1986;64:267–76.

25. Hershenson MB, Kikuchi Y, Loring SH. Relative strengths of the chest wall muscles. J Appl Physiol. 1988;65:852–62.
26. Boriek AM, Wilson TA, Rodarte JR. Displacements and strains in the costal diaphragm of the dog. J Appl Physiol. 1994;76:223–9.
27. Boriek AM, Kelly NG, Rodarte JR, Wilson TA. Biaxial constitutive relations for the passive canine diaphragm. J Appl Physiol. 2000;89:2187–90.
28. Angelillo M, Boriek AM, Rodarte JR, Wilson TA. Theory of diaphragm structure and shape. J Appl Physiol. 1997;83:1486–91.
29. Wait JL, Saworn D, Poole DC. Diaphragm thickness heterogeneity at functional residual capacity and total lung capacity. J Appl Physiol. 1995;78:1030–6.
30. Boriek AM, Liu S, Rodarte JR. Costal diaphragm curvature in the dog. J Appl Physiol. 1993;75:527–33.
31. Road J, Newman S, Derenne JP, Grassino A. In vivo length-force relationship of canine diaphragm. J Appl Physiol. 1985;58:1646–53.
32. Boriek AM, Black B, Hubmayr R, Wilson TA. Length and curvature of the dog diaphragm. J Appl Physiol. 2006;101:794–8.
33. De Troyer A. The action of the canine diaphragm on the lower ribs depends on activation. J Appl Physiol. 2011;111:1266–71.
34. Estenne M, De Troyer A. Relationship between respiratory muscle electromyogram and rib cage motion in tetraplegia. Am Rev Respir Dis. 1985;132:53–9.
35. Mortola JP, Sant' Ambrogio G. Motion of the rib cage and the abdomen in tetraplegic patients. Clin Sci Mol Med. 1978;54:25–32.
36. De Troyer A, Wilson TA. Action of the isolated canine diaphragm on the lower ribs at high lung volumes. J Appl Physiol. 2014;592:4481–91.
37. De Troyer A, Wilson TA. Effect of acute inflation on the mechanics of the inspiratory muscles. J Appl Physiol. 2009;107:315–23.
38. Legrand A, Ninane V, De Troyer A. Mechanical advantage of sternomastoid and scalene muscles in dogs. J Appl Physiol. 1997;82:1517–22.
39. Legrand A, Schneider E, Gevenois PA, De Troyer A. Respiratory effects of the scalene and sternomastoid muscles in humans. J Appl Physiol. 2003;94:1467–72.
40. Leevers AM, Road JD. Mechanical response to hyperinflation of the two abdominal muscle layers. J Appl Physiol. 1989;66:2189–95.
41. De Troyer A, Wilson TA. The parasternal and external intercostal muscles drive the ribs differently. J Appl Physiol. 2000;523:799–806.
42. Konno K, Mead J. Measurement of the separate volume changes of the rib cage and abdomen during breathing. J Appl Physiol. 1967;22:407–22.
43. Loring SH, Mead J. Action of the diaphragm on the rib cage inferred from a force-balance analysis. J Appl Physiol. 1982;53:756–60.
44. Warner DO, Krayer S, Rehder K, Ritman R. Chest wall motion during spontaneous breathing and mechanical ventilation in dogs. J Appl Physiol. 1989;66:1179–89.
45. Wilson TA, De Troyer A. Effects of insertional and appositional forces of the canine diaphragm on the lower ribs. J Physiol. 2013;591:3539–48.
46. Agostoni E, Mognoni P, Torri G, Agostoni A. Static features of the passive rib cage and diaphragm-abdomen. J Appl Physiol. 1965;20:1187–93.
47. Jiang TX, Demedts M, DeCramer M. Mechanical coupling of upper and lower canine rib cages and its functional significance. J Appl Physiol. 1988;64:620–6.
48. D'Angelo E, Bellemare F. Electrical and mechanical output of the inspiratory muscles in anesthetized dogs. Respir Physiol. 1990;79:177–94.
49. Knight H, Petroll WM, Rochester D. Relationships between abdominal and diaphragmatic volume displacements. J Appl Physiol. 1991;71:565–72.
50. Mead J, Loring SH. Analysis of volume displacement and length changes of the diaphragm during breathing. J Appl Physiol. 1982;53:750–5.

51. Ward ME, Ward JW, Macklem PT. Analysis of human chest wall motion using a two-compartment rib cage model. J Appl Physiol. 1992;72:1338–47.
52. Wilson TA. Compartmental models of the chest wall and the origin of Hoover's sign. Respir Physiol Neurobiol. 2015;210:23–9.
53. Wilson TA, Angelillo M, Legrand A, De Troyer A. Muscle kinematics for minimum work of breathing. J Appl Physiol. 1999;87:554–60.
54. Butler JP, Huang J, Loring SH, Lai-Fook SJ, Wang PM, Wilson TA. Model for a pump that drives circulation of pleural fluid. J Appl Physiol. 1995;78:23–9.
55. Wang PM, Lai-Fook SJ. Upward flow of pleural liquid near lobar margins due to cardiogenic motion. J Appl Physiol. 1992;73:2314–9.

Chapter 3
Flow and Gas Transport

Abstract Weibel's symmetrical bifurcation model of the bronchial tree provides the basis for analyzing respiratory flows. Flow is turbulent in the central airways and laminar in the peripheral airways, and for quiet breathing, most of the pressure drop occurs in the central airways. For higher frequency forced oscillatory flow, lung impedance is frequency-dependent if lung elastance or resistance is non-uniform. For higher frequencies, resonances occur. Expiratory flow is limited by the wave-speed condition at higher lung volumes and by viscous flow limitation at lower lung volumes. Maximum expiratory flow is a strong function of lung volume, and the maximum expiratory flow-volume curve changes with disease. During inspiration, inspired gas is carried to the periphery by convection in the larger airways and diffusion in the periphery, and a stationary front with a large gradient in oxygen concentration is established at the boundary between the two regions. Regional ventilation of the parenchyma is markedly non-uniform.

3.1 The Bronchial Tree

Weibel's monograph on lung anatomy was published in 1963 [1]. In this book, Weibel describes morphometric methods for extracting significant information from measurements made on micrographs of fixed lung tissue, he describes the architecture of the alveolar ducts, alveoli, and capillary bed in detail, and he lists the results of systematic measurements of the geometry of the bronchial tree. He describes the bronchial tree as an irregular dichotomously branching tree of ~23 generations. The generations are classified as belonging to the conducting zone, generations 0–16, starting with the trachea; the transitory zone, consisting of the respiratory bronchioles of generations 17–19; and the respiratory zone, consisting of the alveolar ducts and terminal alveolar sacs of generations 20–23. He also lists the dimensions of the airways in an equivalent symmetrically bifurcating tree model. The diameters (d_n) for this model are plotted vs. generation number (n) in Fig. 3.1. Except for the first few generations, the length to diameter ratio is ~3, independent of n. This model has provided the basis for subsequent modeling of flow and gas transport in the airways.

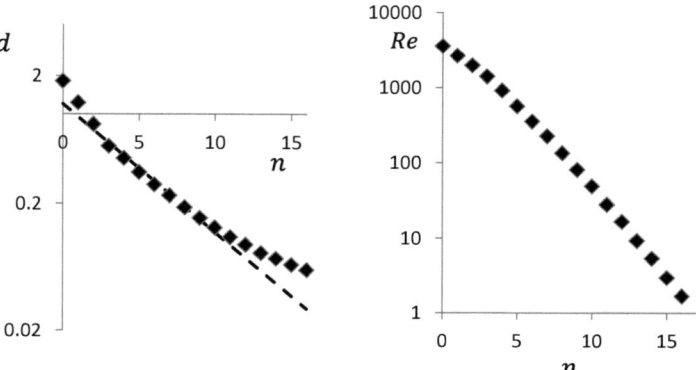

Fig. 3.1 Airway diameter (d) (*left panel*) and Reynolds number (Re) (*right panel*) vs. generation number (n)

Weibel pointed out that the values of d_n for the conducting airways are fit well by the following equation, shown by the dashed line in Fig. 3.1.

$$d_n = d_o/2^{n/3} \quad \text{or} \quad d_{n+1} = d_n/2^{1/3} \qquad (3.1)$$

In hematology, this relation is known as Murray's Law. It can be obtained from a teleological argument [2, 3]. The total metabolic cost to the organism is assumed to be the sum of the cost of maintaining the volume of fluid (blood) or filling the volume with fluid (air) and the cost of viscous dissipation in the flow. Equation (3.1) describes the diameter ratio for which this total cost is minimum.

The geometries of the bronchial trees of dogs and rodents are not well-described as dichotomous branching trees. In these animals, a large tapering central airway with side branches runs nearly the length of each lobe. The side branches lead to asymmetric dichotomously branching sub-trees.

3.2 Flow

Flow in pipes is not a simple phenomenon. The nature of the flow depends on the Reynolds number (Re), a dimensionless number that describes the relative importance of inertial and viscous stresses in the flow; $Re = \rho \cdot \bar{u} \cdot d/\mu$, where $\bar{u}$ is the mean velocity in the pipe, and ρ and μ are the density and viscosity of the fluid; for air, $\rho = 1 \cdot 10^{-3}\,\text{gm/cm}^3$, $\mu = 2 \cdot 10^{-4}\,\text{dyn/cm}^2$. For $Re < 100$, viscous stresses dominate, and the flow is laminar. Fluid particles move in straight lines parallel to the tube axis, and the fluid velocity (u) is a parabolic function of the radial coordinate r: $u = 2 \cdot \bar{u} \cdot \left[1 - \frac{r^2}{(d/2)^2}\right]$ The pressure gradient (dP/dx) in this flow is given by Eq. (3.2) where $\dot{V}$ is the volume flow rate.

3.2 Flow

$$\frac{dP}{dx} = \frac{32 \cdot \mu \cdot \bar{u}}{d^2} = \frac{128 \cdot \mu \cdot \dot{V}}{\pi \cdot d^4} \tag{3.2}$$

For $Re > 2000$, the flow is turbulent. Turbulent flows are chaotic; the fluid velocity at any point is unsteady with components in the transverse directions as well as in the axial direction. The time average velocity is nearly uniform across the pipe with a sharp decrease to zero at the wall. The pressure gradient is given by Eq. (3.3), where the coefficient f is the friction factor.

$$\frac{dP}{dx} = f \cdot \frac{\rho \cdot \bar{u}^2}{2 \cdot d} = f \cdot \frac{8 \cdot \rho \cdot \dot{V}^2}{\pi^2 \cdot d^5} \tag{3.3}$$

The value of f depends on Re and wall roughness. Values of f are given graphically in the Moody diagram. For $Re = 2000$ for a smooth walled tube, $f = 0.05$.

In the intermediate range, $100 < Re < 2000$, turbulent regions are generated by constrictions or bends in the pipe or appear spontaneously and dissipate as they are carried downstream. The pressure drop along the pipe fluctuates even for constant flow.

The values of Re as a function of n for a flow of 1 L/s are shown in the right panel of Fig. 3.2. In generations 0–2, flow is turbulent, for $n = 3-8$, flow is mixed, and for $n > 8$, flow is laminar.

Pedley et al. [4] applied their expertise in fluid mechanics to calculate the dissipative pressure drop (ΔP) in the bronchial tree. For a flow of 1 L/s, they calculated that $\Delta P = 0.7$ cm H_2O. Ninety percent of this pressure drop occurs in the first eight generations of the tree and the pressure drop in the peripheral airways is quite small. Their calculated pressure drop increases in proportion to $\dot{V}^{2/3}$.

Since it was introduced in 1915, Rohrer's equation, given by Eq. (3.4), has been used to describe the pressure loss in the airways. This contains terms like those in Eqs. (3.2 and 3.3), and one would expect a_1 to be proportional to fluid viscosity and a_2 to be proportional to fluid density, but the coefficients in this equation must be determined empirically.

$$\frac{dP}{dx} = a_1 \cdot \dot{V} + a_2 \cdot \dot{V}^2 \tag{3.4}$$

Reynolds and Lee [5] measured the pressure drop for expiratory flow through a model of three generations of the bronchial tree and fit the following equation to the data.

$$\frac{dP}{dx} = 3.2 \cdot \frac{128 \cdot \mu \cdot \dot{V}}{\pi \cdot d^4} + 0.13 \cdot \frac{8 \cdot \rho \cdot \dot{V}^2}{\pi^2 \cdot d^5} \tag{3.5}$$

Thus, Eq. (3.5) provides the values of the coefficients in Rohrer's equation. The first term is the same as Eq. (3.2), but multiplied by 3.2, apparently because of the bifurcations in the tube, and the second is the same as Eq. (3.3) with $f = 0.13$.

In addition to the dissipative pressure losses described above, pressure changes are required to accelerate the flow where the cross-sectional area of the pipe decreases and flow speed increases This pressure gradient is given by Eq. (3.6), where A is cross-sectional area.

$$\frac{dP}{dx} = -\rho \cdot \bar{u} \cdot \frac{d\bar{u}}{dx} = \frac{\rho \cdot \dot{V}^2}{A^3} \cdot \frac{dA}{dx} \qquad (3.6)$$

The integral of Eq. (3.6) is the Bernoulli equation, $P + \frac{1}{2} \cdot \rho \cdot \bar{u}^2 = P + \frac{1}{2} \cdot \rho \cdot \frac{\dot{V}^2}{A^2} = const$. For inspiratory flow, a pressure drop from ambient air to the trachea is required to accelerate the flow. Similarly, during expiration, a pressure drop from alveolar pressure to airway lumen pressure is required. For quiet breathing, the dynamic pressure, $\frac{1}{2} \cdot \rho \cdot \bar{u}^2$, is small, ~0.1 cm H_2O.

3.2.1 Higher Frequency Oscillatory Flows

Dogs and other animals use panting to regulate temperature, and panting is within the repertoire of respiratory maneuvers of humans. A curious feature of panting is that adequate alveolar ventilation is maintained although tidal volume is smaller than the volume of the dead space. Turbulent mixing and dispersion is sufficient to transport blood gases to and from the respiratory zone [6]. This turbulent mixing allows high frequency ventilation to be used to diminish the magnitude of the displacements in surgery fields.

Fig. 3.2 Two-compartment model with different resistances (R) and elastances (E) for the two branches

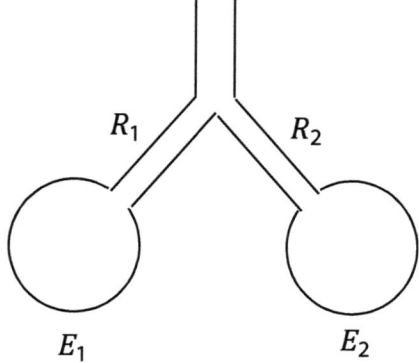

Forced oscillations at higher than normal breathing frequencies are used as diagnostic techniques. In this method, a small amplitude pressure or flow oscillation is imposed at the mouth and the magnitude and phase of the resulting volume or pressure oscillation is measured. The ratio of pressure to flow is respiratory system impedance.

Otis et al. [7] analyzed the impedance of a simple two-compartment model shown in Fig. 3.2. For passive deflation, alveolar pressure and volume for each compartment follow an exponential decrease with time constants E_i/R_i. For an oscillating pressure applied to the airway opening at low frequencies, for which the resistive pressure loss is small, the volume expansions of the compartments are inversely proportional to E_i. This is the distribution of volumes for which the elastic energy stored in the system is minimum for a given total volume, and thus, the distribution for which the elastance of the system is minimum. At high frequencies, for which the resistive losses are large, the volume expansions are inversely proportional to R_i. This is the distribution for which the energy dissipation is minimum for a given total volume oscillation. If the time constants of the compartments are different, the distribution of the volume expansion shifts with increasing frequency, and the elastance of the system increases and the resistance decreases. Thus, a frequency-dependence of elastance and resistance implies an inhomogeneity of time constants within the system. In the usual test, pressure oscillations are applied at the mouth with frequencies ranging from 0.1 to 1 Hz. In normals, the changes of elastance and resistance are negligible. In subjects with asthma (different resistances for different pathways) or emphysema (different elastances and resistances for different pathways), elastance and resistance change by a factor of 2 or more.

The plot of total airway area of the airways in the Weibel model as a function of distance from the larynx is shown in Fig. 3.3. This area distribution is like a tube with nearly constant cross-sectional area (A) of 2.5 cm^2 and length (L) of 22 cm terminated by a bell with a rapid increase in area. For oscillatory flows, a pressure gradient along the tube is required to provide the acceleration of the mass of the air with density ρ in the tube. Thus, the air in the tube provides an inertance (I) of $\rho \cdot L/A$. The system is like a mass-spring system with the air in the airways providing the mass and the elastance of the parenchyma providing the spring. For $I = 10^{-2}$ cm H$_2$O/(L·s^2) and $E = E_L = 4$ cm H$_2$O/L, the resonant frequency (f) for the lungs is $f = (E/I)^{1/2}/2\pi = 3$ Hz. For the respiratory system, the dynamic elastance of the chest wall is in series with the elastance of the lung, the total elastance is >4 times E_L, and $f = 8$ Hz [8].

At even higher frequencies, the pressure differences along the tube combined with the compressibility of air form the familiar resonant system of a tube open at both ends. The resonant frequency for this tube, $f = c/(2L)$, where c is the speed of sound in air, is 80 Hz. The airways form an internal resonator like the external resonators provided by the bodies of wind instruments. This internal resonator is an essential component of the mechanism that excites the vibrations of the vocal cords during speech [9] just as the external resonators of the musical instruments are

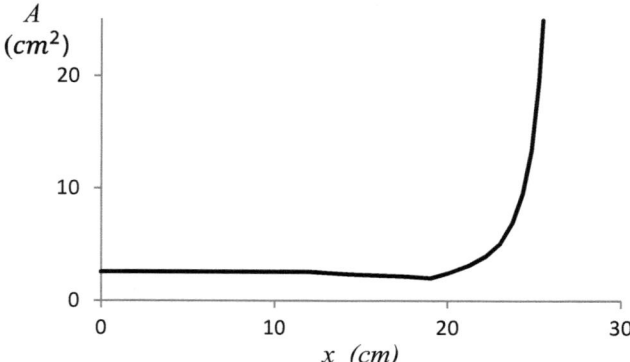

Fig. 3.3 Total cross-sectional area of the airways (A) vs. distance from the larynx (x)

essential to the mechanism that excites the vibrations of the lips of a horn player or the reed of a woodwind. It should also be noted that for frequencies higher than ~50 Hz, most of the signal introduced by oscillations imposed at the mouth is reflected by the rapid area increase at the distal end of the tube. This limits the information about the periphery of the lung that can be obtained by measuring impedance for frequencies higher than 50 Hz.

3.3 Expiratory Flow Limitation

In 1951, Daman [10] reported that flow resistance increased with increasing effort during expiration, and he speculated that the increased resistance was the result of dynamic compression of the airways. That is, gas pressure outside the airway equals alveolar pressure whereas pressure in the lumen of the airway is lower than alveolar pressure because of the pressure drop along the airway from the periphery, and this pressure difference compresses the airway. Fry et al. [11] measured lung volume, pleural pressure and flow for a series of expirations with different expiratory efforts and made three-dimensional graphs of flow vs. the two variables, lung volume and pleural pressure. They found that the line of intersection of the surface of this graph with a plane at a given lung volume showed flow rising with increasing pleural pressure up to a critical pressure, $P^*_{pl}(V_L)$, above which flow was independent of P_{pl}. Hyatt et al. [12] recognized that a two dimension plot of flow vs. expired volume for forced expiratory efforts (1) displayed all the information about flow for $P_{pl} > P^*_{pl}$ (2) could be obtained without the invasive measurement of P_{pl} and (3) showed marked differences between normal subjects and patients with pulmonary disease. This work initiated two decades of research on the characteristics of the flow-volume curve and the search for the mechanism that limits flow.

3.3 Expiratory Flow Limitation

The mechanism for expiratory flow limitation in the lungs is analogous to flow limitation in a rocket engine. In the 1960s and 1970s, the focus of aeronautical engineering was rocket development. A major feature of rockets is the limitation of flow through the nozzle of the rocket. For plenum pressures greater than a critical value, flow is limited by the flow for which the flow speed at the point of minimum area of the nozzle, denoted the choke point, equals the speed of sound at that point. Two engineers working in Jere Mead's laboratory recognized the analogy between flow limitation in a rocket nozzle and expiratory flow limitation [13]. They recognized that the pertinent critical speed in the airways is not the speed of sound in the gas, but the speed of propagation of a small pressure disturbance in a compliant tube. This speed was familiar as the wave speed of the pulse in the arteries. Wave speed (c) in a compliant tube is given by Eq. (3.7), where A is the cross-sectional area of the tube, ρ is the density of the fluid in the tube, P_{tm} is transmural pressure, and dA/dP_{tm} is the compliance of the tube.

$$c = \sqrt{A/\rho \cdot (dA/dP_{tm})} \qquad (3.7)$$

Wave speed depends on the mechanical properties of the airways. The point in the airways where gas velocity equals wave speed is the flow-limiting-site or choke point, and maximum flow ($\dot{V}_{max}$) equals $c \cdot A$ at that point.

This theory of expiratory flow limitation can be demonstrated both graphically and analytically. The graphical demonstration is the following. Representative plots of the total area (A) of all parallel airways for generations (n) 3–6 vs. P_{tm} in humans are shown in the left panel of Fig. 3.4. In both dogs [14] and humans [15], airway area reaches a maximum at $P_{tm} = 10 - 12 \, cm \, H_2O$. The maximum area of individual airways decreases with increasing n, but total area increases with n. Airway compliance at $P_{tm} = 0$ also increases with increasing n. Wave speed can be calculated from these curves as a function of n and P_{tm}, but this is not sufficient information to calculate maximum flow. Both flow speed and P_{tm} at every point in the airways must be known in order to identify the flow for which flow speed equals wave speed at some point in the airways. Note that peri-bronchial pressure (P_{pb}) is the difference between alveolar gas pressure and the tensile stress imposed by parenchymal attachments: $P_{pb} = P_A - P_{tp} = P_{pl}$. Thus, $P_{tm} = P_l - P_{pb} = P_l - P_{pl} = P'_l$, where P_l is gas pressure in the lumen of the airway and prime denotes pressure relative to pleural pressure.

Pressure in the airways (P_l) equals alveolar pressure in the periphery and decreases along the airways due to dissipative pressure losses (ΔP_{diss}) and convective acceleration as described by the Bernoulli equation.

$$P_l = P_A - \Delta P_{diss} - \frac{1}{2}\rho \frac{\dot{V}^2}{A^2} \qquad (3.8)$$

Substituting $P_A = P_{tp} + P_{pl}$ in Eq. (3.8) yields the following equation

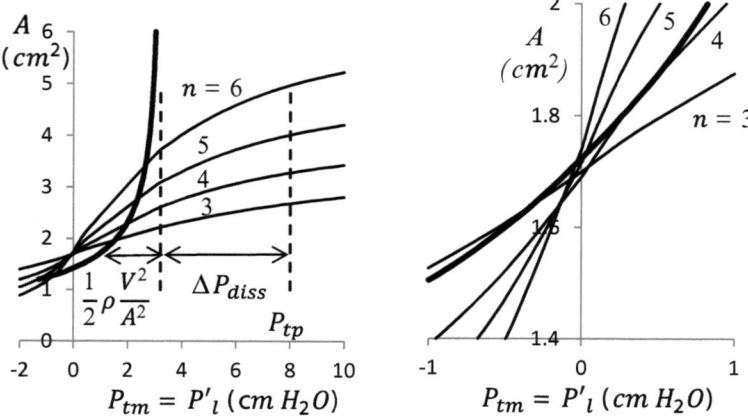

Fig. 3.4 *Left panel*: Area (A) of all parallel airways in a given generation (n) vs. transmural (P_{tm}) or lumen pressure relative to pleural pressure (P'_l) for $n = 3-6$ (*light lines*). Plot of Eq. (3.9) for $P_{tp} = 8$ cm H_2O and $\dot{V} < \dot{V}_{max}$ (*heavy line*). *Right panel*: Magnified view of area-pressure curves near point of tangency of flow line for $\dot{V}_{max}$ and area-pressure curve for $n = 4$

$$P'_l = P_{tp} - \Delta P_{diss} - \frac{1}{2}\rho\frac{\dot{V}^2}{A^2} \qquad (3.9)$$

If it is assumed that ΔP_{diss} occurs primarily upstream of the flow limiting site and that the pressure decrease near the flow limiting site is dominated by the Bernoulli term. Eq. (3.9), for given values of P_{tp} and $\dot{V}$, is a relation between P'_l and A that can be plotted on the same axes as those in Fig. 3.4. This plot is shown by the heavy line in Fig. 3.4. The values of P'_l and A in each generation for this flow are given by the intersection of the flow line with the airway area lines. It can be seen that for higher values of $\dot{V}$, the flow line shifts to the left, and for some flow, the flow line is tangent to the airway area curve for some generation. This is the maximum flow ($\dot{V}_{max}$) for this value of P_{tp}; no solution exists for $\dot{V} > \dot{V}_{max}$. A close-up of the airway area and flow curve for $\dot{V}_{max}$ is shown in the right panel of Fig. 3.4. The flow curve is tangent to the area curve for generation 4. At the point of tangency, the slopes of the two curves are equal. Setting the derivative with respect to A of the airway area curve equal to the derivative with respect to A of the flow curve yields the following equation.

$$dP_{tm}/dA = \rho\dot{V}^2/A^3 \qquad (3.10)$$

This is identical to the wave speed condition. This graphical analysis of flow limitation has been verified by data on airway mechanics, pressure, and flow in excised human lungs [15].

3.3 Expiratory Flow Limitation

The graphical analysis displays several features of maximum flow limitation. First, P'_l at the site of flow limitation is near zero. Second, the intersection of the flow curve with the area curve for generation 3 is to the left of the flow limiting site. For maximum flow, the pressure drop to the flow limiting site is independent of the driving pressure P_{pl} greater than the critical pressure that is required to generate $\dot{V}_{max}$. For driving pressures greater than the critical pressure, additional pressure decreases occur due to compression of airways downstream from the flow limiting site. Third, with decreasing lung volume and decreasing Ptp, the point of tangency shifts to higher n; the choke point moves peripherally as lung volume decreases.

The analytical demonstration is obtained by differentiating Eq. (3.9) with respect to x where x is the coordinate along the airways in the direction of flow.

$$(dP'_l/dx) = -\rho \cdot (\dot{V}^2/A^3) \cdot (dA/dx) - (d\Delta P_{diss}/dx) \tag{3.11}$$

By substituting $(dA/dP'_l) \cdot (dP'_l/dx)$ for (dA/dx) and rearranging terms, Eq. (3.11) can be put in the following form.

$$(dP'_l/dx) = (d\Delta P_{diss}/dx)/[1 - (\dot{V}^2/c^2 \cdot A^2)].$$

Thus, as $\dot{V}/A$ approaches c, the dissipative pressure drop is amplified by dynamic airway compression, and the total pressure drop is unbounded.

These two descriptions of flow limitation are appropriate for higher lung volumes for which convective pressure decreases are significant. For lower volumes for which viscous losses dominate, a different model is needed. This model is a single tube of length L with uniform mechanical properties and with $P'_l = P_{tp}$ at the upstream end and a variable pressure, P_2, at the downstream end. The pressure distribution is assumed to be given by the Poiseuille law, Eq. (3.12), where a is a numerical constant, and μ is the coefficient of viscosity.

$$(dP'_l/dx) = a \cdot \mu \cdot (\dot{V}/A^2) \tag{3.12}$$

Multiplying both sides by $A^2 dx$ yields the following equation.

$$A^2 dP'_l = -a \cdot \mu \cdot \dot{V} \cdot dx$$

Because A^2 is a function of P'_l and $a \cdot \mu \cdot \dot{V}$ is constant, the two sides can be integrated from the upstream end to the downstream end where $P'_l = P_2$ and $x = L$. This yields the following equation for $\dot{V}$.

$$\dot{V} = \frac{1}{a \cdot \mu \cdot L} \int_{P_2}^{P_{tp}} A^2(P'_l) dP'_l$$

If $A^2(P'_l)$ approaches zero faster than $-1/P'_l$ as $P'_l \to -\infty$, the integral is finite for $P_2 \to -\infty$, and a finite flow is produced by an infinite pressure drop.

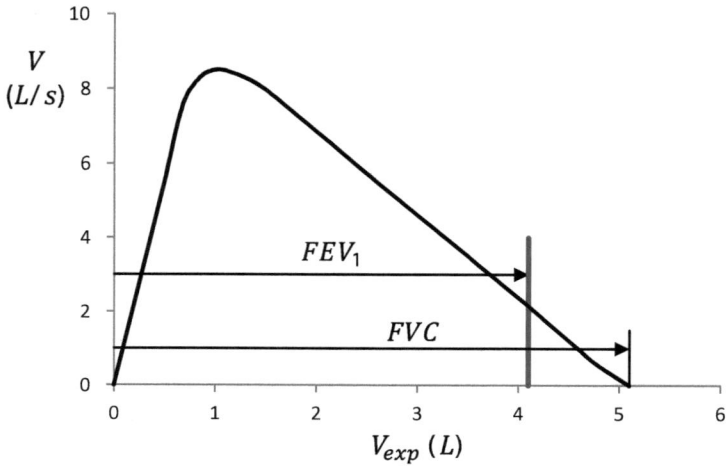

Fig. 3.5 Representative flow-volume curve: flow ($\dot{V}$) vs. expired volume (V_{exp}). Forced vital capacity (*FVC*) is the total volume expired, and FEV_1 is the volume expired in the first second of expiration

To cover the full range of volumes and both flow-limiting mechanisms, a computational model is required [16].

Measurement of the forced expiratory flow-volume curve is the most common test of lung function. In this test, the subject inspires to TLC and flow and expired volume are measured during a forceful expiration. The most accurate measurements are made in a body plethysmograph or "body box" [17]. The subject is enclosed up to the neck in a box, and the change in lung volume is measured by measuring the pressure change of the air in a sealed box or the flow of air into an open box. With this measurement, the contribution of gas compression in the lung is included in the measurement of lung volume. In most cases, flow is measured, and the flow signal is integrated to obtain expired volume.

A representative flow-volume curve for an adult male is shown in Fig. 3.5. The ascending limb of this curve is determined by the rate at which the expiratory muscles are activated and pleural pressure rises. The descending limb is determined by flow limitation. The forced expired volume in one second (FEV_1) or the ratio of FEV_1 to *VC* are used to characterize the curve by a number. In normals, FEV_1 encompasses $\sim 80\%$ of the forced vital capacity (*FVC*).

Variations of the test have been explored in the hope of finding tests that reveal more about lung function. These include the use of gases with different densities and viscosities [18]. At higher lung volumes in normals, the density dependence of maximum flow, $-(\rho/\dot{V}) \cdot (d\dot{V}/d\rho)$, is near its theoretical maximum of 0.5, and viscosity dependence, $-(\mu/\dot{V}) \cdot (d\dot{V}/d\mu)$, is near zero. At lower volumes, density dependence is low and viscosity dependence is near its maximum of 1.0. Partial flow-volume curves in which forced expiration is initiated at volumes below TLC have also been measured. For normals, flow initially spikes slightly above the value for the forced vital capacity maneuver at the same volume, but quickly returns to those values. The overshoot is the result of compression of the airways downstream of the flow-limiting-site [19]. For patients with COPD, initial flow is higher than for the vital capacity maneuver and merges with those values gradually. The higher flows are the result of the contribution of rapidly-emptying regions that would be empty for the VC maneuver. These variations have not been particularly helpful in extending the information that is obtained from the test.

In disease, maximum flows are reduced. In constricted normals and mild asthma, both flows and *FVC* are reduced and the flow-volume line lies parallel to the normal curve, but shifted to the left [20]. This is simply the result of reduced airway caliber [21]. In chronic obstructive pulmonary disease (*COPD*), flows are markedly reduced and the flow-volume curve is frequently scooped with healthier regions emptying faster and regions with greater resistance providing the flows that form the tail of the curve. In *COPD*, parenchymal degeneration reduces the stress applied by parenchymal attachments to the airways and the area-pressure curves are effectively shifted to the right in Fig. 3.4.

The expiratory limb of the flow-volume loop for quiet breathing for normal subjects lies well below limiting flow, but during heavy exercise, part of the limb lies along the maximum flow curve. Patients with mild COPD increase, rather than decrease, their end-expiratory volume during exercise, apparently to decrease their expiration times and increase minute ventilation [22]. With more severe *COPD*, the flow-volume loop for quiet breathing lies along the maximum flow curve, expiratory times are extended, inspiration may begin before expiration to *FRC*, and dynamic hyperinflation occurs.

3.4 Convection and Diffusion

At the beginning of inspiration, ambient air passes through the upper airways (nose and throat), enters the trachea, and flows through the conducting airways toward the periphery, and the interface between the inspired gas and the resident gas in the airways retreats toward the parenchyma. Both convection and diffusion contribute to the transport of oxygen and carbon dioxide in the airways. The relative importance of these two transport mechanisms is described by the dimensionless Peclet number (*Pe*), $\bar{u} \cdot L \cdot D = V \cdot L/A \cdot D$ where L is airway length and D is the coefficient of diffusion, 0.2 cm²/s. *Pe* is large in the trachea and decreases toward the periphery. For $\dot{V} = 250$ mL/s, $Pe \sim 1$ in generation 17. Thus, convection dominates transport in the airways down to the 17th generation and diffusion dominates beyond that point.

In the usual description of transport in the transition region [23], it is assumed that the concentration of a species is uniform across each airway, the total cross-sectional areas of the airways is described as a smooth function of distance, as shown in Fig. 3.6, and the one-dimensional transport equation, Eq. (3.13), is analyzed to obtain the concentration C of a species as a function of axial position x and time t.

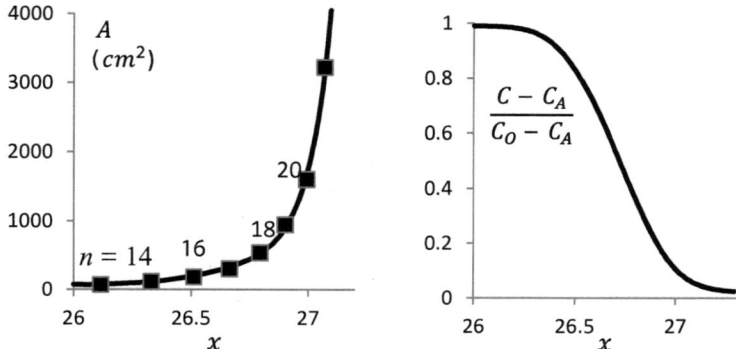

Fig. 3.6 Smooth curve through values of total airway area for Weibel's regular dichotomy model generations, $n = 14\text{--}21$ vs. distance (x) from the entrance to the trachea (*left panel*). Axial dependence of gas concentration C in the stationary front that forms the transition between the concentration C_O of the inspired gas and concentration C_A in the alveolar gas (*right panel*)

$$A(\partial C/\partial t) = -\dot{V}(\partial C/\partial x) + D[\partial A(\partial C/\partial x)/\partial x] \quad (3.13)$$

For steady flow and a sufficiently rapid increase of area with distance, a steady solution exists for boundary conditions, $C = C_O$ upstream and $C = C_A$ downstream, as shown in Fig. 3.6. This steady distribution is described as the stationary front, and any mixing that has occurred upstream of that point by turbulence, secondary flows, and Taylor dispersion is absorbed into the stationary front. The maximum slope of the stationary front occurs at the point where $(\partial A/\partial x) = \dot{V}/D$, and this point lies in generation 17. The bulk of the concentration change occurs in generations 16–19, the respiratory bronchioles. At the end of the 19th generation, C has decreased to within 5 % of the downstream value, and from this point out, diffusion imposes a nearly uniform value of C. The 20th generation is the first generation that is completely alveolated and constitutes the first generation of the alveolar ducts. It should be noted that the flux of gas through the stationary front is constant, equal to $\dot{V} \cdot C_O$; only the mechanism of transport shifts from convection to diffusion.

The conclusion that the concentration of a species is uniform within an acinus, the alveolar volume fed by a terminal bronchial of the 19th generation, is consistent with an estimate of the time for equilibration τ of concentration differences between points separated by a distance a, $\tau = a^2/D$. The size of an acinus is ~0.2 cm, and the value of τ for this scale, ~0.2 s, is small compared to respiratory times. It also follows that inhomogeneity of parenchymal expansion or blood flow at a scale smaller than 0.2 cm has no effect on the efficiency of gas exchange.

The volume of gas upstream of the stationary front, $\sim 200\,\text{cm}^3$, is denoted the dead space volume (V_{DS}) because it does not contribute to alveolar ventilation. For higher values of $\dot{V}$ or for gases with lower diffusivities, the stationary front is shifted slightly to the right and V_{DS} is larger [24].

3.5 Ventilation Distribution

Evidence that alveolar ventilation is not uniform throughout the lung is obtained from the single-breath washout curve. In this test, the subject inspires a breath of pure oxygen and nitrogen concentration in the expired gas is measured during the subsequent expiration. The resulting trace of nitrogen concentration (C), as a fraction of its initial concentration (C_O), vs. expired volume (V_{exp}), is shown in Fig. 3.7. The trace consists of three phases. In phase I the gas in the dead space which contains pure oxygen, is expired. Phase II is the transition phase, and in phase III, the alveolar plateau, mixed alveolar gas reaches the mouth.

Two features of this curve show evidence of nonhomogeneous ventilation. First, the mean concentration of mixed alveolar gas is less than the concentration for ideal mixing. The amount of N_2 in the lung before inspiration, $C_O \cdot (V_{ee} + V_{DS})$, where V_{ee} is the end-expiratory alveolar volume, equals the amount at end inspiration, $C \cdot (V_{ee} + V_T)$. From this it follows that if lung expansion were uniform, $C/C_O = (V_{ee} + V_{DS})/(V_{ee} + V_T)$. Cumming [25] pointed out that the mean concentration in phase III is smaller than the concentration for ideal mixing, and he reported values of the ratio of measured to ideal, denoted the mixing efficiency, of ~92 %. A mixing efficiency of less than 1 implies that some regions of the lung receive more than their share of inspired gas, have lower concentrations of N_2 at end inspiration, and contribute more to the subsequent expiration than other regions.

Second, the alveolar plateau is not flat; N_2 concentration rises steadily over phase III. In his seminal paper on the slope of phase III, Fowler [26] argued that the slope of phase III implied both a spatial and a temporal heterogeneity of ventilation with well-ventilated regions emptying earlier and poorly-ventilated regions later in expiration. Both of the markers of ventilation heterogeneity are enhanced in disease.

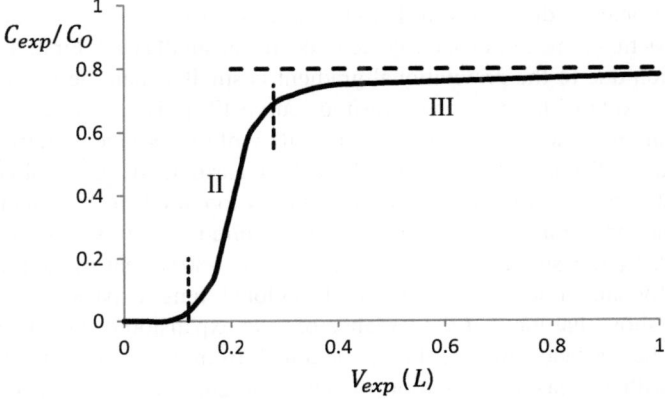

Fig. 3.7 Concentration of N_2 in the expired gas (C_{exp}), as a fraction of its initial concentration (C_O) as a function of expired volume (V_{exp}) after a single breath of pure O_2. The concentration for uniform mixing is shown by the horizontal *dashed line*

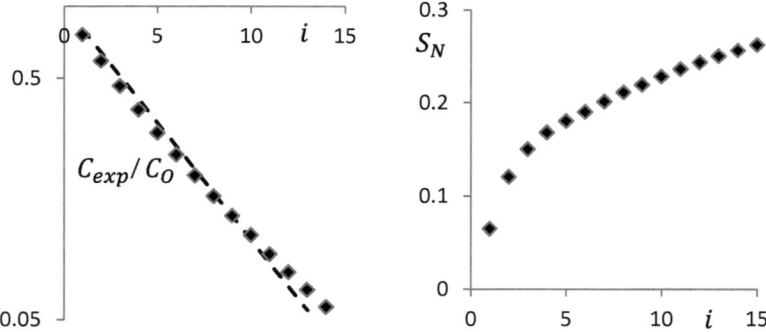

Fig. 3.8 Mean alveolar concentration in expired gas vs. breath number (i) (*left panel*) and normalized slope of Phase III (S_N) vs. i (*right panel*) for multi-breath washout

Additional information about heterogeneous ventilation is obtained from the multi-breath washout test. In this test, the subject continues to breath pure oxygen and mean alveolar concentration and the slope of phase III are measured for all breaths. A representative plot of $C(i)/C_O$ vs. breath number (i) is shown in the semi-log plot on the left of Fig. 3.8. The straight line plot of this function for ideal mixing is shown by the dashed line. The slope of the plot of measured values for small i is steeper than the slope for ideal mixing because well-ventilated regions dominate the signal for early breaths. The slope decreases with increasing i because the concentration in well-ventilated regions becomes small and the signal is dominated by the expired gas from poorly-ventilated regions.

A plot of the normalized slope of phase III ($S_N(i)$), slope divided by ($C_{exp}(i)$), vs. i is shown in the right panel of Fig. 3.8. S_N rises rapidly over the first few breaths and then continues to rise, but more slowly, for subsequent breaths.

Even now, some textbooks state that the source of heterogeneous ventilation is the gravitational gradient of ventilation described in Chap. 1. Considerable evidence shows that most of the heterogeneity occurs at small scale. First, the variance of ventilation due to the gravitational gradient is smaller than the variance that is required to explain the measured washout curve [27]. The variance of regional specific ventilation described in Chap. 2 is <40 % of the variance inferred from the washout curve. Second, the slope of phase III measured for gas expired from an airway that serves a segment of the lung that is too small to be affected by the gravitational gradient is ~60 % of that for expired gas from the whole lung [28]. Third, S_N measured in a gravity-free state is ~80 % that measured on earth [29]. Finally anatomic measurements of regional lung expansion by imaging techniques show that most of the variance of lung expansion occurs at small scale [30–32]. The variance of ventilation obtained from these imaging techniques increases with increasing resolution of the technique, but, to date, the variance obtained from imaging is smaller than that inferred from functional data.

By fitting the predictions of a model of heterogeneous ventilation to the data, the magnitude of the variance of ventilation can be determined [33]. In the model, the

3.5 Ventilation Distribution

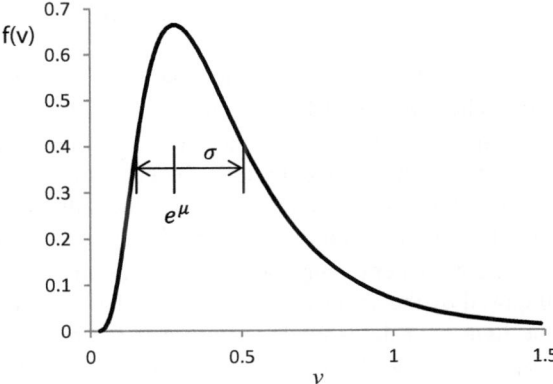

Fig. 3.9 Plot of Eq. (3.14) vs. v on a linear scale. The values of v at one SD above and below the mean of ln v are marked

alveolar volume at end expiration (V_{ee}) is imagined to be divided into a large number of units with equal volumes. The volume increase of each unit during inspiration, divided by its volume at end-expiration, is denoted v. For variables that are intrinsically positive, but have coefficients of variation that are large, the log-normal distribution is useful.

$$f(v) = \frac{1}{\sqrt{2\pi} \cdot \sigma} exp\left[-\frac{(\ln v - \mu)^2}{2 \cdot \sigma^2}\right] \quad (3.14)$$

The value of $f(v)$ describes the fraction of the lung with specific ventilation v. The distribution contains two parameters, the mean value of $\ln v$, denoted μ, and the SD of $\ln v$, denoted σ. The symbol σ has the same meaning as the symbol Log SD that is used in the description of the $\dot{V}_A/\dot{Q}$ distribution [34]. The values of $C_{exp}(i)$ are calculated for this distribution (Appendix), and the values of μ and σ that give the best fit to the observed curve are determined. The value of V_T during the washout is known so that the values of μ and σ obtained from the fit provide the value of V_{ee}. The wash-in of an inert gas, usually He, provides comparable data.

For normal subjects with $V_{ee} = 3$ L and $V_T = 1$ L, the values of the parameters of the distribution are $\mu = -1.28$ and $\sigma = 0.6$. The distribution f for these values is shown vs. v with a linear scale in Fig. 3.9. The peak of the distribution occurs at $v = e^\mu = 0.28$. The value of $\bar{v}$ is higher: $\bar{v} = 0.33$. The heterogeneity of regional ventilation is large. The value of v at $\ln v = \mu + \sigma$ is more than 3 times the value at $\ln v = \mu - \sigma$.

The ventilation distribution is physiologically significant because it is one of the constituents of the ventilation/perfusion ratio [34] distribution that determines the efficiency of gas exchange. Because $ln(v/q) = \ln v - \ln q$, where q is the blood flow per unit volume at end expiration, it follows that the variance of $\ln v/q$, denoted $\sigma^2_{v/q}$, is related to the variances of $\ln v$ and $\ln q$ by the following equation, where ρ is the coefficient of correlation between $\ln v$ and $\ln q$.

$$\sigma_{v/q}^2 = \sigma_v^2 - 2\rho \cdot \sigma_v \cdot \sigma_q + \sigma_q^2$$

A study of the wash-in of soluble gases gave the result that $\sigma_q \sim 0.8$, and $\rho \sim 0.85$ [35]. Thus, $\sigma_{v/q} \sim 0.4$, in agreement with the value obtained from data on gas exchange, namely the differences between the partial pressures of soluble gases in expired gas and in mixed arterial blood [34].

Severinghouse et al. [36] occluded the pulmonary artery of one lung and found that the elastance and resistance of that lung increased, and ventilation was shifted to the contralateral lung, thereby ameliorating the nonuniformity of $\dot{V}_A/\dot{Q}$ that was imposed by the occlusion of the artery. The increase of elastance and resistance in the lung with the occluded artery was the result of smooth muscle activation in response to the decrease in the concentration of CO_2 in that lung. There is no evidence that this mechanism functions at a smaller scale [37].

The slope of phase III has little physiological significance, but it has received considerable experimental [38–40] and theoretical attention [41–43]. Paiva and Engel [42] proposed a mechanism for the temporal inhomogeneities that generate the slope of phase III. They hypothesized asymmetries in the geometry of the branches of terminal bronchiole so that the diffusional resistances of the two branches are different. The branch with the lower diffusional resistance is better ventilated and empties earlier. This model has been elaborated by Verbanck and colleagues [41, 43]. The modeling is limited to the analysis of a single bifurcation or a single acinus. For large asymmetries, the model predictions match the observed values of S_N for the first few breaths, but the predictions fall below the observations for later breaths. Also, the washout curve predicted by these models does not match the experimental curve. A model that describes a purely convective mechanism has also been proposed [44]. In this model, differences in alveolar size and cable stiffness produce differences in regional compliance, and surface tension relaxation causes phase differences between the volume oscillations of regions with different specific expansion. The assumptions in both of these model have not been confirmed experimentally.

Thus, this section is different from the others in this book. It provides a phenomenological description of ventilation heterogeneity without describing the mechanisms, because these mechanisms have not been determined.

Appendix

It is assumed that each unit receives gas from the dead space in proportion to its volume expansion. At the time during inspiration when the gas originally in the dead space has entered the alveolar volume, the volume of the unit is $(1 + \alpha \cdot v)$ where $\alpha = V_{DS}/V_T$ and the concentration of N_2 in the unit is C_O. At end inspiration, the volume of the unit is $(1 + v)$. Therefore, the concentration of N_2 in the unit at end-inspiration of the first breath $(C(v, 1))$ is given by Eq. (3.15).

$$C(v, 1) = \frac{(1 + \alpha \cdot v)}{(1 + v)} \qquad (3.15)$$

During the next expiration. expired gas from all units is assumed to be well mixed, and expired gas concentration is the sum of the contributions from all units.

$$C_{exp}(1) = \frac{V_{ee}}{V_T} \cdot \int C(v, 1) \cdot v \cdot f(v) \cdot dlnv \qquad (3.16)$$

In subsequent breaths (breath i), the concentration in the dead space is $C_{exp}(i-1)$ and the mass balance for the unit yields the following equation for $C(v, i)$.

$$C(v, i) = \frac{[1 + \alpha \cdot v \cdot C_{exp}(i - 1)]}{(1 + v)}. \qquad (3.17)$$

$C_{exp}(i)$ is then calculated from the equation corresponding to Eq. (3.16).

References

1. Weibel ER. Morphometry of the human lung. New York: Academic; 1963.
2. Murray CD. The physiological principle of minimum work: I the vascular system and the cost of blood volume. Proc Natl Acad Sci U S A. 1926;12:207–14.
3. Wilson TA. The design of the bronchial tree. Nature. 1967;213:668–9.
4. Pedley TJ, Schroter RC, Sudlow MF. The prediction of pressure drop and variation of resistance within the human bronchial airways. Respir Physiol. 1970;9:387–405.
5. Reynolds DB, Lee SJ. Modeling study of the pressure-flow relationship of the bronchial tree. Fed Proc Fed Am Soc Exp Biol. 1979;38:1444.
6. Fredberg JJ. Augmented diffusion in the airways can support pulmonary gas exchange. J Appl Physiol. 1980;49:232–8.
7. Otis AB, McKerrow CB, Bartlett RA, Mead J, McIlroy MB, Selverstone NJ, Radford Jr EP. Mechanical factors in distribution of pulmonary ventilation. J Appl Physiol. 1956;8:427–34.
8. Michaelson ED, Grassman ED, Peters WR. Pulmonary mechanics by spectral analysis of forced random noise. J Clin Invest. 1975;56:1210–30.
9. Gupta V, Wilson TA, Beavers GS. A model for vocal cord excitation. J Acoust Soc Am. 1974;54:1607–17.
10. Damon H. Mechanics of airflow in health and disease. J Clin Invest. 1951;30:1175–90.
11. Fry DL, Ebert RV, Stead WW, Brown CC. The mechanics of pulmonary ventilation in normal subjects and in patients with emphysema. Am J Med. 1954;16:80–97.
12. Hyatt RE, Schilder DP, Fry DL. Relationship between maximum expiratory flow and degree of lung inflation. J Appl Physiol. 1958;13:331–6.
13. Dawson SV, Elliott EA. Wave-speed limitation on expiratory flow—a unifying concept. J Appl Physiol. 1977;43:498–515.
14. Sittipong R, Hyatt RE. Static mechanical behavior of bronchi in excised dog lung. J Appl Physiol. 1974;37:201–6.
15. Hyatt RE, Wilson TA, Bar-Yishay E. Prediction of maximal expiratory flow in excised human lungs. J Appl Physiol Respir Environ Exerc Physiol. 1980;48:991–8.

16. Lambert RK, Wilson TA, Hyatt RE, Rodarte JR. A computational model for expiratory flow. J Appl Physiol. 1982;52:44–56.
17. Mead J. Volume displacement body plethysmography for respiratory measurements in human subjects. J Appl Physiol. 1960;15:736–40.
18. Staats BA, Wilson TA, Lai-Fook SJ, Rodarte JR, Hyatt RE. Viscosity and density dependence during maximal flow in man. J Appl Physiol. 1980;48:313–9.
19. Knudson RJ, Mead J, Knudson DE. Contribution of airway collapse to supramaximal flow. J Appl Physiol. 1974;36:653–67.
20. Moore BJ, Hilliam CC, Verburgt LM, Wiggs BR, Vedal S, Pare PD. Shape and position of the complete dose response curve for inhaled methacholine in normal subjects. Am J Respir Crit Care Med. 1996;154:642–8.
21. Lambert RK, Wilson TA. Smooth muscle dynamics and maximal expiratory flow in asthma. J Appl Physiol. 2005;99:1885–90.
22. Babb TG, Viggiano R, Hurley B, Staats B, Rodarte JR. Effect of mild-to-moderate airflow limitation on exercise capacity. J Appl Physiol. 1991;70:223–30.
23. Wilson TA, Lin KH. Convection and diffusion in the airways and the design of the bronchial tree. In: Bouhuys A, editor. Airway dynamics, physiology and pharmacology. Springfield: Charles Thomas; 1970.
24. Berdine GG, Johnson JE, Dale D, Lehr JL. Inspiratory flow rate and dead space in dogs. J Appl Physiol. 1990;68:1228–32.
25. Cumming G. Gas mixing efficiency in the human lung. Respir Phsyiol. 1967;2:213–24.
26. Fowler WS. Lung function studies III. Uneven pulmonary ventilation on normal subjects and in patients with pulmonary disease. J Appl Physiol. 1949;2:283–99.
27. Piiper J, Scheid P. Respiration: alveolar gas exchange. Annu Rev Physiol. 1971;33:131–54.
28. Engel LA, Utz G, Wood LDH, Utz G, Macklem PT. Ventilation distribution in anatomical lung units. J Appl Physiol. 1974;37:194–200.
29. Prisk GK, Elliott AR, Guy HJ, Verbanck S, Paiva M, West JB. Multiple-breath washin of helium and sulfur hexafluoride in sustained microgravity. J Appl Physiol. 1998;84:244–52.
30. Hoffman EA, Chon D. Computed tomography studies of lung ventilation and perfusion. Proc Am Thorac Soc. 2005;2:492–8.
31. Hubmayr RD, Walters BJ, Chevalier PA, Rodarte JR, Olson LE. Topographical distribution of regional lung volume in anesthetized dogs. J Appl Physiol. 1983;54:1048–56.
32. Musch G, Layfield JDH, Harris RS, Melo MFV, Winkler T, Callahan RJ, Fischman AJ, Venegas JG. Topographical distribution of pulmonary perfusion and ventilation, assessed by PET in supine and prone humans. J Appl Physiol. 2002;93:1842–51.
33. Beck KC, Wilson TA. Variance of ventilation during exercise. J Appl Physiol. 2001;90:2151–6.
34. West JB, Wagner PD. Ventilation-perfusion relationships. In: Crystal RG, West JB, editors. The lung: scientific foundations. New York: Raven; 1991.
35. Beck KC, Johnson BD, Olson TP, Wilson TA. Ventilation-perfusion distribution in normal subjects. J Appl Physiol. 2012;113:872–7.
36. Severinghouse JW, Swenson EW, Finley TN, Lategola MT, Williams J. Unilateral hypoventilation produced in dogs by occluding one pulmonary artery. J Appl Physiol. 1961;16:53–60.
37. Altemeier WA, Robertson HT, McKinney S, Glenny RW. Pulmonary embolization causes hypoxemia by redistributing regional blood flow without changing ventilation. J Appl Physiol. 1998;85:2337–43.
38. Crawford ABH, Makowska M, Engel LA. Effect of tidal volume on ventilation maldistribution. Respir Phsyiol. 1986;66:11–25.
39. Crawford ABH, Makowska M, Kelly S, Engel LA. Effect of breathholding on ventilation maldistribution during tidal breathing in normal subjects. J Appl Physiol. 1986;61:2108–15.
40. Crawford ABH, Makowska M, Paiva M, Engel LA. Convection- and diffusion-dependent ventilation maldistribution in normal subjects. J Appl Phsyiol (1985). 1985;59:838–46.

References

41. Dutrieue B, Vanholsbeeck F, Verbanck S, Paiva M. A human acinar structure for simulation of realistic alveolar plateau slopes. J Appl Physiol. 2000;89:1859–67.
42. Paiva M, Engel LA. Pulmonary interdependence of gas transport. J Appl Physiol. 1979;47:296–305.
43. Verbanck S, Paiva M. Model simulations of gas mixing and ventilation distribution in the human lung. J Appl Physiol. 1990;69:2269–79.
44. Wilson TA. Parenchymal mechanics, gas mixing, and the slope of phase III. J Appl Physiol. 2013;115:64–70.

Index

A
Abdominal muscles, 33
Acinus, 3, 54, 58
Airways, 1, 3, 7, 9–13, 21, 25, 27, 31, 33–35, 43–51, 53–54, 56
Alveolar walls, 3–6, 10, 11

B
Bernoulli equation, 46, 49
Bronchial tree, 3, 43–45
Bulk modulus, 11, 12, 14

C
Campbell diagram, 37, 38
Choke point, 49, 51
Compartmental models, 33–38
Convection, 53–54

D
Dead space, 46, 54–55, 58
Diaphragm, 8, 9, 20, 28–37
Diffusion, 53–54

E
Elastance, 1, 3, 5, 6, 8, 9, 12, 35, 46, 47
Expiratory flow limitation, 48–53

G
Gas transport, 43–58
Gravitational deformation, 8, 13

H
Hamberger model, 27

I
Impedance, 47, 48
Intercostal muscles, 21–28, 32
Interdependence, 10, 12, 13

M
Mechanical advantage, 25, 26, 30, 33
Mixing efficiency, 55
Murray's law, 44

O
Oscillatory flow, 46–48

P
Peclet number, 53
Phase III, 55–56, 58
Pleural space, 33, 37–38
Pressure-volume, 1, 2, 6, 7, 11, 14, 21

R
Resistance
 airway, 7
 tissue, 7
Resonance, 43
Reynolds number, 44
Rib cage, 5, 9, 21–28

S
Shear modulus, 11, 14
Smooth muscle, 12
Stationary front, 54
Surfactant, surface tension, 4, 5

T
Tissue
 contribution to recoil, 6
 resistance, 7

V
Ventilation, 8, 9, 13, 46, 53–58

W
Wave speed, 49, 50
Work of breathing, 37

Z
Zone of apposition, 29–32, 36

MIX
Papier aus verantwortungsvollen Quellen
Paper from responsible sources
FSC® C105338

If you have any concerns about our products,
you can contact us on
ProductSafety@springernature.com

In case Publisher is established outside the EU,
the EU authorized representative is:
**Springer Nature Customer Service Center GmbH
Europaplatz 3, 69115 Heidelberg, Germany**

Printed by Libri Plureos GmbH
in Hamburg, Germany